W9-AXR-741

TROPICALS

TROPICALS

GORDON COURTRIGHT

TIMBER PRESS
Portland, Oregon

ISBN 0-88192-098-3
Printed in Hong Kong

Timber Press
9999 SW Wilshire
Portland, Oregon 97225

Library of Congress Cataloging-in-Publication Data

Courtright, Gordon.
    Tropicals / Gordon Courtright.
        p.   cm.
    Bibliography: p.
    Includes index.
    ISBN 0-88192-098-3
    1. Tropical plants--Dictionaries. 2. Plants, Ornamental-
-Dictionaries. 3. Tropical plants--Pictorial works. 4. Plants,
Ornamental--Pictorial works.   I. Title.
SB407.C684 1988
    635.9'52--dc19   88-23524
                                                               CIP

CONTENTS

This book is intended to be a visual plant dictionary.

TO EACH OF YOU
WHO WISHES TO KNOW
BY NAME
YOUR FRIENDS IN THE GARDEN

ACKNOWLEDGMENTS

It gives me a great deal of pleasure to acknowledge those individuals whose assistance made this book a reality.

—My wife, Addi, for her everlasting tolerance and support.
—John Bryan, Horticultural Consultant, who kept things straight on a number of occasions.
—Celia Wittenber, my typist, who not only checked me botanically but also made me abide by the "King's English."
—Julian Herman, Landscape Architect, North Hollywood, California.
—Gary Powell, Haleina, Hawaii
—The very cooperative staffs of the following botanic gardens:
   Durban Botanical Garden, Durban, South Africa
   Fairchild Botanical Gardens, Miami, Florida
   Foster Botanical Gardens, Hawaii
   Huntington Botanical Gardens, Los Angeles, California
   Kirstenbosch Botanical Gardens, Cape Town, South Africa
   Los Angeles Botanical Gardens, Los Angeles, California
   Lyons Botanical Gardens, Hawaii
   Selby Botanical Gardens, Sarasota, Florida
   Waimea Botanical Gardens, Hawaii

To all, a warm thank you.

INTRODUCTION

This book is intended to be a visual dictionary of the plants usually seen when one visits a tropical area in the United States or elsewhere in the world.

The idea for this book came to mind 25 years ago, on my first trip to Hawaii. Having been in the nursery business for more than 20 years at the time, I found that I recognized and knew the names of most of these plants, as they are grown in subtropical latitudes of the mainland from California to Florida. However, a number of new plants and flowers still confronted me, and, unfortunately, I could find no single book to help me identify them and learn their names.

In retirement, I have at last found the time to take the pictures and prepare this book to satisfy my desire to make things a little easier for the people who follow.

The names, both botanic and common, used for ornamental plants are a problem for anyone who buys, sells, or uses them. One of the difficulties a horticultural writer must deal with is keeping up with the ongoing changes in botanic names, a problem shared with botanists who trouble to keep up with the scientific botanic literature published in most countries. However, nomenclature committees, working worldwide, have been and are trying to clear up this confusion in botanic plant names, so I use them to organize this book. It must be pointed out that many of the names used in the 1930s and '40s have been changed, so I have included the older synonyms for the benefit of the reader who has not the time to keep current with the botanic nomenclature.

I also have added the English language common names in the descriptions when known—some plants have none, some as many as five—and the number would be increased significantly if those in other languages were included. The same common name, however, is used for different species from one country to the next. In locating these disparate common names I have consulted the horticultural literature of Australia, New Zealand, South Africa, and Europe, as well as North America.

Despite this confusion about names—which, unfortunately, will continue to remain as long as research and discovery are pursued—it is my earnest desire that you enjoy using this book as much as I have enjoyed compiling it.

One final note. After I published my first book, many people asked about the pictures—who took them, what kind of camera was used, etc. In this book, more than 90% were taken by myself, mainly with a Minolta XGl with a 50-mm lens, the last few with a Minolta 7000. All the photos, except about a dozen, were taken with a hand-held camera, and all were taken in the U.S.A.

TEMPERATURE RATINGS

Zone 10	40° to 30°
Zone 9	30° to 20°
Zone 8	20° to 10°
Zone 7	10° to 0°
Zone 6	0° to —10°
Zone 5	—10° to —20°
Zone 4	—20° to —30°
Zone 3	—30° to —40°

Temperatures suggested in this book are approximate. The growing conditions, for example, a warm, late fall, often will keep the plants from hardening off and a sudden cold snap has been known to freeze plants 20° above the normal freezing point.

BOTANIC INDEX

BOTANIC	COMMON NAME	PLATE NUMBER
Brahea edulis	Guadalupe Fan Palm	343
Brassaia actinophylla	Octopus Tree	190
	Queensland Umbrella Tree	
Breynia disticha 'Roseo-picta'	Snow Bush	21
Bromeliads		22–22N
Brownea capitella	Brownea	191, 191A
Brugmansia suaveolens	Angels Trumpet	23
	Maikoa	
Brugmansia versicolor	Apricot Moonflower	24, 24A
Brunfelsia nitida	Lady-of-the-Night	25
Brunfelsia pauciflora 'Floribunda'	Yesterday-Today-and-Tomorrow	26, 26A
Brunfelsia pauciflora 'Macrantha'		27, 27A
Butea monosperma	Flame-of-the-Forest	192, 192A
	Parrot Tree	
Butia capitata	Hardy Blue Cocos	344
	Jelly Palm	
Caesalpinia pulcherrima	Barbados Flower Fence	28, 28A
	Dwarf Poinciana	
	Pride of Barbados	
Calathea insignis	Rattlesnake Plant	29, 29A
Calliandra haematocephala	Pink Powderpuff	30, 30A
Calliandra portoricensis		31
Calliandra tweedii	Brazilian Flame Bush	32
	Mexican Flame Bush	
Callistemon citrinus	Bottlebrush	33
Callistemon viminalis	Weeping Bottlebrush	193, 193A
Calodendrum capense	Cape Chestnut	194, 194A
Calotropis gigantea	Crown Flower	34
	Mudar	
Campsis × tagliabuana 'Madame Galen'	Trumpet Creeper	291
Carica papaya	Fruta Bomba	35
	Papaya	
	Pawpaw	
Carissa macrocarpa	Natal Plum	36
Caryota cumingii	Fishtail Palm	345
Caryota mitis	Burmese Fan Palm	346
	Clustered Fishtail Palm	
Caryota no	Fishtail Palm	347
Cassia alata	Candlestick Senna	37, 37A
	Christmas Candle	
Cassia excelsa	Cassia	195
	Crown of Gold	
Cassia fistula	Golden Shower	196
	Indian Laburnum	
Cassia javanica	Apple Blossom	197
	Pink-and-White Shower Tree	
Cassia javanica × C. fistula	Rainbow Shower Tree	198, 198A
Cassia surattensis	Kolomona	38, 38A
Casuarina equisetifolia	Australian Pine	199
	Beefwood	
	Horsetail Tree	
Ceiba pentandra	Kapok	200
	Silk-Cotton Tree	
Cerbera manghas		201, 201A
Chamaerops humilis	Hair Palm	348
	Mediterranean Fan Palm	
Chorisia speciosa	Floss Silk Tree	202, 202A
Chrysalidocarpus lutescens	Areca Palm	349
	Butterfly Palm	
Chrysophyllum cainito	Caimito	203
	Star Apple	

BOTANIC	COMMON NAME	PLATE NUMBER
Cibotium chamissoi	Hapu'u-ii	39
	Hawaiian Tree Fern	
Clerodendrum macrostegium		40
Clerodendrum speciosissimum	Java Glorybower	41, 41A
Clerodendrum splendens		42
Clerodendrum thomsonae	Bleeding Heart Vine	292
	Glory Bower	
Clivia miniata	Kaffir Lily	43
Clusia rosea	Autograph Tree	204, 204A
	Scottish Attorney	
Clytostoma callistegioides	Lavender Trumpet Vine	293, 293A
	Love-charm	
Coccoloba uvifera	Kino	205
	Platterleaf	
	Sea Grape	
Coccothrinax alta	Silver Palm	350
Cocculus laurifolius	Laurel-leaved Snailseed	44
Cochlospermum vitifolium	Buttercup Tree	206–206B
	Wild Cotton	
Cocos nucifera	Coconut Palm	351, 351A
Codiaeum variegatum	Croton	45–45C
Coffea arabica	Arabian Coffee	46, 46A
Colocasia esculenta	Elephant's Ear	47
	Kalo	
	Taro	
Combretum microphyllum	Burning Bush	48
	Flame Creeper	
Congea tomentosa	Pink Shower Orchid	294, 294A
	Shower-of-Orchids	
Copernicia hospita	Cuban Wax Palm	352
Cordia boissiere	Anacahutta	207, 207A
Cordia sebestena	Geiger Tree	208, 208A
Cordyline australis atropurpurea	Bronze Dracaena	49
Cordyline indivisa	Dracaena Palm	50
Cordyline stricta	Palm Lily	51
Cordyline terminalis	Good Luck Plant	52
	Ti Plant	
Corypha umbraculifera	Talipot Palm	353, 353A
Costus speciosus	Crape Ginger	53
	Malay Ginger	
Costus spicatus	Shampoo Ginger	54
Couroupita guianensis	Cannonball Tree	209–209B
Crinum asiaticum	Poison Bulb Lily	55
	Spider Lily	
Crinum augustum 'Queen Emma'	Queen Emma Lily	56, 56A
Crotalaria agatiflora	Canarybird Bush	57
Cryptostegia grandiflora	Rubber Vine	295
Cupaniopsis anacardioides	Carrotwood	210
	Tuckeroo	
Cuphea hyssopifolia	Elphin Herb	58
	False Heather	
Cuphea ignea	Cigar Flower	59
	Firecracker Plant	
Curcuma cocana	Hidden Lily	60
Cycas circinalis	Fern Palm	354
	Queen Cycas	
	Sago Palm	
Cycas revoluta	Japanese Fern Palm	355, 355A
	Sago Palm	
Cyperus alternifolius	Umbrella Plant	61
Cyperus papyrus	Bulrush	62
	Egyptian Paper Plant	
	Papyrus	
Delonix regia	Flamboyant	211–211B
	Flame Tree	
	Royal Poinciana	

SHRUBS

1   *Acalypha hispida*

2   *Acalypha wilkesiana*

Acalypha hispida   Chenille Plant
                                                      Red Cats-tail

Family: Euphorbiaceae
Origin: East Indies

1   A shrub to 10 ft. with 4–6 in. long, bright green leaves, 8 in. long, and dark red flowers, forming velvety tails up to 18 in. long. Likes a shaded area but needs some sun to develop the flowers. Flowers most of the year.   Zone 10

Acalypha wilkesiana   Copper-leaf
                                                   Jacob's Coat
                                            Match-Me-If-You-Can

Family: Euphorbiaceae
Origin: Fiji

2   This monoecious shrub, shown here with hanging male flowers, grows to about 15 ft., with bronzy green foliage mottled copper-red. The flower spikes are about 8 in. long and ¼ in. wide. Best grown in full sun. It will drop its leaves during a cold snap.   Zone 10

Acalypha wilkesiana 'Godseffiana'   Jacob's Coat

Family: Euphorbiaceae
Origin: Garden hybrid

3   This is one of many cultivars of *A. wilkesiana* with the same common name. It grows to the same size as the species but the leaves are bright green with a yellow edge. Best in full sun for the most colorful foliage. Seen here with upright female flowers.
                                                      Zone 10

3   *Acalypha wilkesiana* 'Godseffiana'

3A   *Acalypha wilkesiana* 'Godseffiana'

4   *Acanthus mollis*

6   *Agave attenuata*

Agave attenuata

Family: Agavaceae
Origin: Mexico

6   This succulent has a huge rosette of large, fleshy, gray-green, spineless leaves, 2–3 ft. long and 10 in. wide. The trunk is up to 5 ft. in height. It is conspicuous at flowering time when a spike-like, greenish yellow, 10 ft. inflorescence suddenly appears, later bending over under its own weight.   Zone 10

7   *Ajuga reptans*

Acanthus mollis   Bear's Breech

Family: Acanthaceae
Origin: Mediterranean Region

4   A perennial, with notched, dark green leaves, 3 ft. long and 1 ft. wide. The flowers are spikes, to 5 ft. with individual green-to-purple blossoms along the stem. It dies to the ground in late summer but comes back very strongly after a short rest.   Zone 9

Adhatoda cydoniaefolia   Brazilian Bower Plant

Family: Myoporaceae
Origin: Brazil

5   A shrub to 6 ft. with pointed, oval leaves up to 10 in. long. The flowers, which are white, grow in clusters of many crimson bracts, which almost completely cover the small flowers.   Zone 10

5   *Adhatoda cydoniaefolia*

Ajuga reptans   Bugleweed

Family: Labiatae
Origin: Europe

7   This perennial grows 3–10 in. high and often is used as a ground cover. There are a number of cultivars, with leaf colors ranging from green to purple to variegated while leaf size varies from small to large. All have 3–5 in. flower spikes in summer.   Zone 6

Alocasia macrorrhiza (A. indica)   Ape
Giant Elephant's Ear

Family: Araceae
Origin: Sri Lanka, India,
Malaya, Indonesia
8   An evergreen shrub reaching a height of 16 ft. widely distributed in the tropics. The large, heart-shaped leaves reach 2 ft. or more in width and 2 ft. in length on 3–5 ft. stalks.   Zone 10

8   *Alocasia macrorrhiza*

9   *Aloe arborescens*

Aloe arborescens   Candelabra Plant
Torch Plant
Tree Aloe

Family: Liliaceae
Origin: South Africa
9   A succulent shrub to 10–12 ft. and sometimes of an equal spread. The plant branches at will. From each terminal rosette a 2 ft. flower stem with a dense raceme of 1 in. scarlet flowers appears from December to February.   Zone 9

Aloe barbadensis (A. vera)   Barbados Aloe
Medicinal Aloe

Family: Liliaceae
Origin: Mediterranean
10   A succulent with narrow, upright, fleshy leaves, 1–2 ft. tall and clusters of yellow flowers atop a 3 ft. stalk. A favorite home remedy to treat burns, insect bites, and sunburn.   Zone 8

10   *Aloe barbadensis*

11 *Alpinia purpurata*

11A *Alpinia purpurata*

11B *Alpinia purpurata*

Alpinia purpurata   Red Ginger

Family: Zingiberaceae
Origin: Malay Peninsula

11   One of the very "showy" plants seen on a tropical visit. The foot-long, red plume, however, is not the flower but the bract covering the small, ¼ in., white flowers hidden within. It is widely used as a foundation planting and grown in fields for commercial florists the year round.

A. p. 'Double'; called 'Tahitian Ginger'; the flower is often 6 in. in
    diameter.

A. p. 'Jungle Queen': a light-pink variety.   Zone 10

Alpinia zerumbet (A. speciosa;   Pink Porcelain Lily
Catimbrum speciosum; Languas   Shell Ginger
speciosa)

Family: Zingiberaceae
Origin: China; Japan

12   Grows to 12 ft., usually in large clumps. The lance-shaped leaves are 2 ft. long and 5 in. across. The flowers, white tinged with purple, emerge from slow opening, shell-pink buds on drooping, elongated clusters during the spring and summer.   Zone 10

12 *Alpinia zerumbet*

13   *Alsophila australis*

14   *Alternanthera ficoidea* 'Versicolor'

Alsophila australis (A. cooperi; Cyathea   Australian Tree Fern
cooperi; Sphaeropteris cooperi)

Family: Cyatheaceae
Origin: Australia

13   There are approximately 300 species found in the world. The nursery trade, however, has not been able to agree on which is which, as they are very much alike as a rule. Most grow to a height of 30 ft. in the wild.   Zone 9 to 26°

Alternanthera ficoidea 'Versicolor'   Joy Weed

Family: Amaranthaceae
Origin: Mexico

14   This selection of the native green variety is dwarf, with leaves blotched and veined with yellow-orange and red. It often is grown as an annual.   Zone 7

15   *Angelonia salicariifolia*

Angelonia salicariifolia

Family: Scrophulariaceae
Origin: Tropical America

15   A tropical perennial, to about 2 ft., bearing fragrant, lavender-to-blue flowers in clusters throughout the year.   Zone 10

Anthurium andraeanum   Flamingo Lily

Family: Araceae
Origin: Colombia

16   This is the most popular flower grown in the Hawaiian Islands and is shipped all over the world as a cut flower. There are many colors—red, orange, pink, white—each with a colorful spathe.   Zone 10

Aptenia cordifolia (Mesembryan-   Baby Sun Rose
themum cordifolium)

Family: Aizoaceae
Origin: South Africa

17   A prostrate, succulent plant used mainly as a ground cover. It grows to about 4 in. in height, with a 2 ft. spread. The rose-red flowers occur in early spring and summer.   Zone 10

16   *Anthurium andraeanum*

17   *Aptenia cordifolia*

18   Bambusa oldhamii

19   Banksia ashbeyi

19B   Banksia hookeriana

19A   Banksia grandis

19C   Banksia ericifolia

Bambusa oldhamii (Sinocalamus oldhamii)
Family: Gramineae
Origin: China, Taiwan

Giant Timber Bamboo
Timber Bamboo

18   The poles will grow up to 4 in. in diameter and to 40 ft. in height if given plenty of heat and water. In 20 years' time, it can grow into thick clumps 10 ft. in diameter.   Zone 10

Banksia
Family: Proteaceae
Origin: Australia

19   A fascinating group of Australian plants of the Protea family. In the warmer areas, they are grown for cut flowers, which are shipped all over the world.   Zone 10

20   Beaucarnia recurvata

21   Breynia disticha 'Roseo-picta'

Beaucarnea recurvata (Nolina tuberculata)
Family: Agavaceae
Origin: Texas; Mexico

Bottle Ponytail
Elephant Foot Tree

20   This dracaena-like plant has a slender trunk(s) and a swollen base. It can grow to 30 ft., with leaves as long as 6 ft. An excellent tub plant for deck planting, as it like dry, desert-like conditions, in full sun. It demands well-drained, sandy loam.   Zone 9

*Breynia disticha 'Roseo-picta'
(B. nivosa; Phyllathus nivosa)*
Family: Euphorbiaceae
Origin: South Pacific

Snow Bush

21   A nice shrub, from 3–6 ft., with leaves that are about 2 in. long and are variegated green, white, pink, and red. Propagated by suckers or cuttings that transplant readily. Can be invasive.   Zone 10

BROMELIADS

Family: Bromeliaceae
Origin: Central America and other Tropical Areas
(Pronunciation: BRO-MEEL-EE-AD)

22–22N Hundreds of cultivars are being grown as garden plants all over the world in the warmer areas and as indoor potted plants everywhere. There are more than 50 species and thousands of cultivars. Here are just a few.

22A Bromeliad aechmea fasciata

22B Bromeliad tillslavellata

22 Bromeliad annas comosus

22C Bromeliad aechmea bracteata

22D Bromeliad billbergia

22E Bromeliad vriesea splendens

22F   *Bromeliad gusmania wittmackii*

22G   *Bromeliad gusmania wittmackii* var.

22H   *Bromeliad gusmania amaranus*

22I   *Bromeliad gusmania cv.*

22J   *Bromeliad gusmania sanguinea*

22K   *Bromeliad vriesea pulmonata*

22L   *Bromeliad vriesea marie*

22M   *Bromeliad neoregelia carolinae*

22N   *Bromeliad neoregelia carolinae*

23   *Brugmansia suaveolens*

*Brugmansia suaveolens (Datura   Angels Trumpet
suaveolens)   Maikoa*

Family: Solanaceae
Origin: S. E. Brazil

23   A fast-growing evergreen shrub or small tree to 20 ft., with 10 in. long leaves and an abundance of large, trumpet-shaped, pendant, white flowers from early spring to late fall. The flowers, fruit, and leaves are all poisonous. All parts of the plant contain a strong narcotic, and the leaves sometimes are smoked by the natives as a relief from asthma—but I would advise against it.   Zone 10

24   *Brugmansia versicolor*

Brugmansia versicolor (Datura mollis)   Apricot Moonflower

Family: Solanaceae
Origin: Ecuador

24   A small, graceful tree to 12 ft., with large, 10 in., peach-colored trumpets in great profusion.   Zone 9

24A   *Brugmansia versicolor*

Brunfelsia nitida   Lady-of-the-Night

Family: Solanaceae
Origin: West Indies

25   A free-flowering shrub to 6 ft., with glossy foliage, 2–3 in. long. The flowers, about 3 in. long, have a greenish-white tube, 3–4 in. long, that flares out to ¾ in. in an orange-like color. Fragrant only at night.   Zone 10

25   *Brunfelsia nitida*

26   Brunfelsia pauciflora

Brunfelsia pauciflora 'Floribunda'   Yesterday-Today-
(B. calycina)   and-Tomorrow

Family: Solanaceae
Origin: Brazil

26   An evergreen shrub from 4–6 ft., with foliage about 3 in. long.
The 1 in. wide, fragrant flowers first have deep lavender buds
(yesterday), open to light lavender flowers (today), and then fade
to white (tomorrow).   Zone 10

26A   Brunfelsia pauciflora

27   Brunfelsia pauciflora 'Macrantha'

Brunfelsia pauciflora 'Macrantha'
(B. grandiflora)

Family: Solanaceae
Origin: Venezuela

27   This evergreen shrub has large leaves, up to 8 in., and is more
tender than the 'Floribunda.' The flowers are 2–4 in.
across.   Zone 10

Caesalpinia pulcherrima (Poinciana   Barbados Flower
pulcherrima)   Fence
Family: Leguminosae   Dwarf Poinciana
Origin: West Indies   Pride of Barbados

28   Big clusters of fiery red flowers grow on the ends of this large,
10–12 ft. shrub, which has lacy foliage and prickly branches. It
blooms almost the year around and is widely planted in the
tropics. The plant is evergreen in the mildest climates, and grows
in full sun and is drought- and salt-tolerant. All parts are used
medicinally. There also is a yellow variety, *C. p.* var. *flava*.
                                                              Zone 10

27A   Brunfelsia pauciflora 'Macrantha'

28 Caesalpinia pulcherrima

28A Caesalpinia pulcherrima

29 Calathea insignis

29A Calathea insignis

Calathea insignis   Rattlesnake Plant

Family: Marantaceae
Origin: Mexico to Ecuador

29   This very interesting plant grows to 6 ft., with leaves more than 2 ft. long and 1 ft. wide. The yellow flowers look like the tail of a rattlesnake.   Zone 10

30 Calliandra haematocephala

Calliandra haematocephala   Pink Powderpuff
(C. inaequilatera)

Family: Leguminosae
Origin: Bolivia

30   A spectacular, large evergreen shrub or small tree to 16 ft., covered with large, bright pink, powderpuff balls, which contrast with the rich, green foliage. Blooms summer and fall.   Zone 9

30A Calliandra haematocephala

31 *Calliandra portoricensis*

Calliandra portoricensis
Family: Leguminosae
Origin: Southern Mexico to Panama

31 An evergreen tree or shrub to 20 ft., with 1½ in. diameter flowers, which have white stamens tipped with pink or red and tiny leaves. It flowers freely for most of the year.   Zone 10

32 *Calliandra tweedii*

Calliandra tweedii (Inga pulcherrima)   Brazilian Flame Bush
Family: Leguminosae   Mexican Flame Bush
Origin: Brazil

32 The leaves on this 8 ft. evergreen shrub are lacy and fern-like. Each branch bears a medium-sized head of fluffy, vivid scarlet stamens shaped like pom-poms. Blooms all spring and summer.   Zone 10

33 *Callistemon citrinus*

Callistemon citrinus (C. lanceolatus)   Bottlebrush
Family: Myrtaceae
Origin: Australia

33 A large evergreen shrub or small tree to about 20 ft. The flowers look like bunches of 4 in., bright red brushes in May and June, with a lesser number of flowers almost all of the year.   Zone 9

34 *Calotropis gigantea*

Calotropis gigantea   Crown Flower
Family: Asclepiadaceae   Mudar
Origin: India, Indonesia

34 An evergreen shrub or small tree to 15 ft. It is a member of the Milkweed family. The milky juice is toxic to some people but butterflies love it. The leaves, roots, and bark are used for medicinal purposes. The flowers range from white through lavender.   Zone 10

35 *Carica papaya*

Carica papaya   Fruta Bomba
Family: Caricaceae   Papaya
Origin: Tropical America   Pawpaw

35 This is one of the favorite backyard fruit trees in Hawaii and other tropical areas. The large leaves, up to 2 ft. across, are found at the top of a 10–20 ft., non-branched trunk, with the delicious fruit growing the year around just under the crown of the leaves. This evergreen tree lives only about 15 years.   Zone 10

36 *Carissa macrocarpa*

37 *Cassia alata*

Carissa macrocarpa (C. grandiflora)   Natal Plum

Family: Apocynaceae
Origin: Tropical Africa

36 An evergreen widely cultivated in the tropics and subtropics as foundation plantings and hedges, grows to 10 ft. or more but there are low-growing, compact cultivars available for landscaping. It has white, star-like flowers, followed by tart, edible fruit.   Zone 10

Cassia alata   Candlestick Senna
Christmas Candle

Family: Leguminosae
Origin: Tropical America

37 This evergreen shrub to 8 ft. has erect "candles" of golden yellow flowers at the ends of its branches. With a little care, it can be a handsome, brightly colored plant, but one can have problems with seedlings.   Zone 8

Cassia surattensis (C. glauca)   Kolomona

Family: Leguminosae
Origin: Tropical Asia to Australia

38 This is a plant that blooms almost every day of the year. The dark green, compound foliage makes an excellent background for the bright yellow flowers and long, brown seedpods. It grows wild in dry areas to 8–10 ft. and even a greater spread. It is evergreen.   Zone 10

37A *Cassia alata*

38 *Cassia surattensis*

38A *Cassia surattensis*

Cibotium chamissoi (C. splendens)   Hapu'u-ii
                                                        Hawaiian Tree Fern

Family:Dicksoniaceae
Origin: Hawaii

39   This fern develops a trunk up to 25 ft., with broad, divided fronds to 12 ft. long, which are found at the top of the trunk. At one time nearly 400,000 acres of these tree ferns grew in the Hawaiian Islands. Before the turn of the century they were gathered and dried and used for filling material for mattresses and pillows, until a better material was found.   Zone 10

39   *Cibotium chamissoi*

Clerodendrum macrostegium

Family: Verbenaceae
Origin: Philippines

40   A bushy shrub to 8 ft., with 6 in., pointed leaves and small clusters of pink flowers on the end of each branch. Likes partial shade.   Zone 10

40   *Clerodendrum macrostegium*

Clerodendrum speciosissimum   Java Glorybower
(C. fallax)

Family: Verbenaceae
Origin: Java

41   The large, heart-shaped leaves are thick and velvety and about 1 ft. in length, making this shrub a very nice addition to the garden. The profusion of reddish flower clusters grow to 12 in. long and about 6 in. above the leaves. The plant itself reaches about 12 ft.   Zone 10

41 · *Clerodendrum speciosissimum*

41A   *Clerodendrum speciosissimum*

42   Clerodendrum splendens

Clerodendrum splendens
Family: Verbenaceae
Origin: Tropical Africa
42   A twining, evergreen vine-shrub with 6 in. long leaves. Clusters of bright red, 1 in. flowers in summer. The plant grows well in the shade and spreads rapidly by root suckers.   Zone 10

43   Clivia miniata

Clivia miniata (Imantophyllum   Kaffir Lily
miniatum)
Family: Amarylliadaceae
Origin: South Africa
43   An excellent plant, for both the shaded garden and as a pot plant. The thick, strap-shaped, evergreen leaves are arranged opposite each other on a fleshy stock from which the flower stalk appears, each with up to 20 orange flowers, produced in winter and very early spring.   Zone 10

Cocculus laurifolius   Laurel-leaved
                                                Snailseed

Family: Menispermaceae
Origin: Southern Japan to the Himalayas
44   An evergreen shrub to 15 ft. The shiny, leathery, 6 in. leaves are a ''black'' deep green and often used in flower arrangements.   Zone 9

44   Cocculus laurifolius

45   Codiaeum variegatum

45A   Codiaeum variegatum

Codiaeum variegatum (Croton pictum)   Croton
Family: Euphorbiaceae
Origin: Malaysia

45   An evergreen, to 6 ft. or more, extensively grown in the tropics (or under glass in colder areas) for the colored, ornamental foliage. Better color can be had if planted in sunny locations. It is widely used in tropical areas for hedges or as specimen plant. There are many cultivars and colors used.   Zone 10

45B   Codiaeum variegatum

45C   Codiaeum variegatum

46 *Coffea arabica*

46A *Coffea arabica*

Coffea arabica   Arabian Coffee

Family: Rubiaceae
Origin: Africa

46 An evergreen shrub widely grown in the tropics around the world for the coffee beans and as a house plant elsewhere. It is a shrub to 15 ft. with wavy, pointed, dark green leaves and small, fragrant, white flowers, which are followed by ½ in. berries that turn red when ripe.   Zone 10

Colocasia esculenta (Caladium esculentum)   Elephant's Ear
                              Kalo
                              Taro

Family: Araceae
Origin: Sri Lanka

47 A perennial grown commercially in the tropics. In Hawaii and other Pacific islands, the root is eaten in the form of "poi" and often grown around pools as an ornamental, where it thrives if the roots can get into the water.   Zone 10

Combretum microphyllum   Burning Bush
                              Flame Creeper

Family: Combretaceae
Origin: Mozambique

48 A large, vine-like, rambling evergreen shrub with long sprays of scarlet flowers. It often reaches an enormous size, covering arbors, trellises, and fences.   Zone 10

47 *Colocasia esculenta*

48 *Combretum microphyllum*

49   *Cordyline australis atropurpurea*

50   *Cordyline indivisa*

Cordyline australis atropurpurea   Bronze Dracaena
(Dracaena australis atropurpurea)

Family: Agavaceae
Origin: New Zealand

49   This attractive plant will grow to 15 ft., but it should be cut
back when very young to force a multiple trunk. It likes deep soil
and can grow at the seashore, beside a swimming pool, or in the
desert.   Zone 8

Cordyline stricta (C. congesta; Dracaena   Palm Lily
stricta)

Family: Agavaceae
Origin: Australia

51   Grows to 12 ft. and is used both indoors or outdoors in
heavily shaded locations. The leaves are narrow, very dark green,
and up to 2 ft. long. It has interesting spikes of small, lavender
flowers.   Zone 9

Cordyline indivisa (Dracaena indivisa)   Dracaena Palm

Family: Agavaceae
Origin: New Zealand

50   Grows to 25 ft., with 3–6 ft. long, sword-shaped, drooping
leaves. It has large clusters of fragrant, white flowers. The plant
branches wherever it blooms. In landscaping it usually is asso-
ciated with Spanish architecture.   Zone 8

Cordyline terminalis (Dracaena   Good Luck Plant
terminalis)   Ti Plant (Pronounced
Family: Agavaceae   ''Tea'')
Origin: Asia to Polynesia

52   Various shades of red, from deep maroon to bright pink, as
well as a solid green form. Widely used by florists all over the
world. The natives use the fiber for thatching, cloth, and food for
livestock. The roots are edible. Grows to 10 ft. in height with 5 in.
wide, 2 ft. leaves.   Zone 10

51   *Cordyline stricta*

52   *Cordyline terminalis*

53   *Costus speciosus*   54   *Costus spicatus*   55   *Crinum asiaticum*

Costus speciosus   Crape Ginger
                                       Malay Ginger

Family: Zingiberaceae
Origin: East Indies
53   Grows from 4–10 ft. tall in many parts of Hawaii and tropical America. The flowerheads consist of large, reddish-purple flower bracts with a white corolla and a white lip touched with yellow, but the color is variable.   Zone 10

Costus spicatus (C. cylindricus)   Shampoo Ginger

Family: Zingiberaceae
Origin: West Indies
54   Grows to 6–10 ft. with thick, smooth leaves. The 3 in., round, bright red bracts and small, yellow flowers develop at the ends of the shoots. The bracts remain closed until the flowers are very old.   Zone 10

56   *Crinum augustum* 'Queen Emma'

Crinum asiaticum (C. floridanum)   Poison Bulb Lily
                                        Spider Lily

Family: Amaryllidaceae
Origin: Tropical Asia and Africa
55   This onion-like bulb grows in large clumps. The leaves, to 4 ft., are dark green to purple. White flowers, as many as 12–20, show a touch of maroon and grow on a thick, purple stem that is taller than the leaves.   Zone 9

Crinum augustum 'Queen Emma'   Queen Emma Lily

Family: Amaryllidaceae
Origin: Tropical Asia
56   This widely planted bulb is found in both tropical and subtropical areas. The flowers are white with a greenish tube, often tipped with red. The bulbs are up to 6 in. in diameter and are poisonous, but, after roasting, they are laid on the skin to ease rheumatic pain. In the Far East, leaf juices often are used to relieve earache.   Zone 10

56A   *Crinum augustum* 'Queen Emma'

57   *Crotalaria agatiflora*

58   *Cuphea hyssopifolia*

59   *Cuphea ignea*

Crotalaria agatiflora   Canarybird Bush
Family: Leguminosae
Origin: Africa
57   A fast-growing, evergreen shrub, to 10 ft. and as wide, with gray-green foliage. A dozen or so chartreuse flowers, 1½ in. across on 15 in. spikes, are shaped like a bird's beak, hence the common name.   Zone 9

Cuphea hyssopifolia   Elphin Herb
                                         False Heather

Family: Lythraceae
Origin: Mexico and Guatemala
58   Compact, evergreen shrublet, 2–3 ft. tall, with tiny flowers of pink, purple, or white in the summer.   Zone 9

Cuphea ignea   Cigar Flower
                                  Firecracker Plant

Family: Lythraceae
Origin: Mexico and Guatemala
59   A nice, small, shrubby, evergreen plant to 2 ft., grown in the tropics. The 1 in. flowers are small and tubular, bright red with white tips. Even a small, 2 ft. diameter plant often will bear 1,000 blooms and always be in bloom.   Zone 9

Curcuma cocana   Hidden Lily
Family: Zingiberaceae
Origin: East Indies
60   One of the group of ginger plants that gives us turmeric and curry. It is cultivated in the tropics. The colorful "flower" is hard and waxy, 12–15 in. long, and about 4 in. in diameter.   Zone 10

60   *Curcuma cocana*

61 *Cyperus alternifolius*

62 *Cyperus papyrus*

63 *Dichorisandra thyrsiflora*

64 *Dicksonia antarctica*

65 *Dieffenbachia*

Cyperus alternifolius   Umbrella Plant

Family: Cyperaceae
Origin: Madagascar

61   A reed-like perennial with stems that are headed with fine, feathery foliage, like a palm. It thrives best in a moist spot, so will grow well in shallow water, such as a fish pond. Grows in sun or shade.   Zone 8

Cyperus papyrus   Bulrush (of the Bible)
Egyptian Paper Plant
Papyrus

Family: Cyperaceae
Origin: North Africa

62   Tall, reed-like stems to 8 ft. are topped by an 8 in., brush-like umbel of lacy, thread-like grass. Plant in a wet spot; this is one of the few plants that can be grown in a fish pond. Papyrus, the writing paper of ancient Egypt, was made from thin strips of the pith that were pressed together while still wet.   Zone 9

Dichorisandra thyrsiflora   Blue Ginger

Family: Commelinaceae
Origin: Brazil

63   This blue ginger grows from 3–6 ft. and has smooth, spirally arranged, shiny green leaves. The plant is found in semi-shaded, moist areas with rich soil as it is a greedy feeder. Although commonly called "Blue Ginger," it is not a member of the Ginger family.   Zone 10

Dicksonia antarctica   Tasmanian Tree Fern

Family: Dicksoniaceae
Origin: Tasmania; Australia

64   A beautiful, garden tree fern in the right place. As many as 50 6 ft. fronds will develop from the woody trunk. After a few years, however, the trunk becomes too tall, eventually up to 35+ ft., for the small garden.   Zone 10

Dieffenbachia spp.   Dumb Cane

Family: Araceae
Origin: Tropical America

65   There are 30 species, most growing to about 6 ft., with leaves varying in color from dark green to yellow or chartreuse. They will not stand any frost. The acrid sap burns the mouth and throat and can leave a rash on the hands and arms similar to poison oak.   Zone 10

66   *Dietes bicolor*

67   *Dietes iridioides*

Dietes bicolor (Moraea bicolor)   African Iris

Family: Iridaceae
Origin: South Africa

66   An evergreen perennial with sword-shaped leaves up to 2½ ft. Lemon-yellow flowers with a brown spot. Blooms only when the sun shines and closes up as night falls.   Zone 9

Dietes iridioides (Moraea iridioides)   African Iris
                                                            Butterfly Iris

Family: Iridaceae
Origin: South Africa

67   This evergreen perennial blooms only when the sun shines and closes at night. It grows to 24 in. in height, has narrow, iris-like foliage, and 3 in., iris-like, white flowers marked with blue and yellow.   Zone 8

Dracaena deremensis 'Janet Craig'

Family: Agavaceae
Origin: Tropical Africa

68   This plant often is used in heavily shaded areas in tropical gardens or as an indoor pot plant elsewhere. It has broad, deep green leaves and grows up to 12 ft.   Zone 10

68   *Dracaena deremensis* 'Janet Craig'

69   *Dracaena fragrans* 'Massangeana'

Dracaena fragrans 'Massangeana'   Corn Plant

Family: Agavaceae
Origin: West Africa

69   An upright plant to 20 ft., with heavy, ribbon-like leaves 2 ft. or more in length and 4 in. wide, which are green with a broad, yellow stripe down the center. It does not like wind. Widely grown in colder areas as a house plant.   Zone 10

70 *Dracaena marginata*

71 *Duranta repens*

72 *Echium fastuosum*

73 *Ensete ventricosum* 'Maurellii'

Dracaena marginata

Family: Agavaceae
Origin: Madagascar

70   This interesting plant grows to about 12 ft. in a lazy, attractive manner, twisting and turning. It can be cut at any point and rerooted. In colder areas, it very often is used as a twisted, specimen house plant.   Zone 10

Duranta repens (D. ellisia; D. plumieri)   Golden Dewdrop
                                                      Pigeon Berry
                                                      Sky Flower

Family: Verbenaceae
Origin: Florida to Brazil

71   An evergreen shrub or tree to 15 ft. with drooping or trailing branches, lacy foliage, and small bunches of lavender flowers. Clusters of the flowers, as well as the orange berries, hang on for most of the year. The fruit is poisonous to humans, and the plant itself is poisonous to cattle.   Zone 10

Echium fastuosum   Pride of Madeira

Family: Boraginaceae
Origin: Canary Islands

72   A large, 4–5 ft., shrubby perennial which grows as clumps of gray-green leaves in a rounded mound. Large, delphinium-like spikes of blue-purple flowers, up to 3 ft. above the foliage, give it a bold effect. It is good in a dry location, as well as at the seashore.   Zone 9

Ensete ventricosum 'Maurellii' (*Musa*   Red Abyssinian
'Maurellii')   Banana
                                                   Red Banana

Family: Musaceae
Origin: Garden Origin

73   This 10 ft., palm-like tropical plant has huge, broad foliage, the stems and veins of which are a burgundy-red.   Zone 10

74   *Eranthemum pulchellum*

Eranthemum pulchellum (E. nervosum)   Blue Sage
Family: Acanthaceae
Origin: India
74   An evergreen shrub to 4 ft., with leaves up to 8 in. long and lovely, tubular, lavender flowers growing in small clusters in the spring. The plant can "escape" in tropical regions and become a pest.   Zone 10

75   *Eugenia uniflora*

Eugenia uniflora   Pitanga
                                                    Surinam Cherry
Family: Myrtaceae
Origin: Brazil
75   An evergreen shrub or small tree to 20 ft. and equally as broad. In southern Florida it is widely grown as a 3–6 ft. hedge. Although closely clipped, it continues to fruit. The fruit, the size of a small tomato, often is eaten as a spicy, tart fruit, although it sometimes is sweet. It is excellent either fresh or preserved.   Zone 10

75A   *Eugenia uniflora*

Euphorbia cotinifolia (E. scotana)   Bronze Euphorbia
                                              Fire-on-the-Mountain
                                              Red Spurge
Family: Euphorbiaceae
Origin: Venezuela and Mexico
76   This bushy, deciduous shrub or small tree grows to a height of 20 ft. or more. The foliage is outstanding, ranging from purplish-green to blood-red, with light colors on the underside. The flowers are small, white, and bell-shaped. The plant, especially the sap, can produce a rash and blisters on the skin.   Zone 10

76   *Euphorbia cotinifolia*

Euphorbia leucocephala   Pascuita
                                      White Lace Euphorbia

Family: Euphorbiaceae
Origin: Southern Mexico to El Salvador
77 A slender shrub or small tree to 10 ft., with slender, narrow leaves, which are a fresh green. The bush forms a cloud of tiny white flowers in the autumn. It needs a well-drained, good soil and a warm climate. The sap can cause a rash to those who are allergic.   Zone 10

Euphorbia milii (E. bojeri)   Christ Thorn
                                      Crown of Thorns

Family: Euphorbiaceae
Origin: Madagascar
78 A very spiny, succulent shrub to 4 ft. with slender, twisting, thorny branches. It flowers the year around with clustered, minute flowers, flanked by a pair of red, petal-like, ½ in. bracts. It loves full sun and requires very little moisture, so it makes a fine pot plant for the deck. The sap is toxic.   Zone 9

Euphorbia pulcherrima (Poinsettia pulcherrima)   Christmas Flower
                                      Mexican Flameleaf
                                      Poinsettia

Family: Euphorbiaceae
Origin: Mexico; Central America
79 A winter-flowering shrub to 10–15 ft. The actual flowers are small and carried in clusters in the center of the bright red bracts. In temperate climates, it is cultivated in hothouses as the "Christmas Flower." In recent years many cultivars have appeared—pink, cream, and variegated, as well as larger individual flower bracts.   Zone 10

Euphorbia tirucalli   Finger Tree
                                      Milkbush
                                      Pencil Tree

Family: Euphorbiaceae
Origin: Tropical and Southern Africa
80 A dioecious, succulent, spineless tree to 30 ft., with only minute leaves and soon deciduous. It is grown in the garden in the tropics but indoors in colder areas.   Zone 10

' 77 *Euphorbia leucocephala*

78 *Euphorbia milii*

79 *Euphorbia pulcherrima*

80 *Euphorbia tirucalli*

81   *Fatsia japonica*

82   *Feijoa sellowiana*

83   *Galphimia glauca*

Fatsia japonica (Aralia japonica;   Japanese Aralia
A. sieboldi)

Family: Araliaceae
Origin: Japan

81   Grown for the bold, evergreen effect of the large, 16 in. foliage. Looks best when 4 or more plants are grown together. It can be grown both indoors and outdoors and will grow to 20 ft. but is best when kept pruned to a smaller size.   Zone 8

Feijoa sellowiana   Pineapple Guava

Family: Myrtaceae
Origin: South America

82   Evergreen shrub to 20 ft. but will stand pruning and can be kept clipped to a hedge. It has glossy, gray-green foliage and whitish flowers with many crimson stamens. The fruit is 2–3 in. long, pineapple-scented, with seeds surrounded by pulp similar to an avocado. It is grown in both tropical and semitropical areas for the fruit, which is eaten raw or preserved as jam or jelly.
                                                        Zone 9 to 22°

Galphimia glauca (G. nitida; Thryallis   Mexican Gold Bush
glauca)   Shower of Gold

Family: Malpighiaceae
Origin: Mexico

83   An evergreen shrub growing to about 8 ft. It has light gray, glossy leaves, about 2 in. long, borne on stems covered with reddish hairs. The 1 in. long, bright yellow flowers are borne in dense clusters, almost covering the bush. In tropical areas, it is an excellent foundation plant, which blooms the year around.
                                                        Zone 10

83A   *Galphimia glauca*

84 *Gardenia jasminoides*

85 *Ginoria glabra*

Gardenia jasminoides (G. florida; G. grandiflora) — Cape Jasmine / Common Gardenia

Family: Rubiaceae
Origin: China

84 This excellent, evergreen bush grows to 6 ft. It requires an acid soil and a warm, sunny location. It is used extensively in fragrant flower corsages. Zone 8 to 18°

Ginoria glabra

Family: Lythraceae
Origin: Central America

85 A spreading, deciduous shrub to about 6 ft. It is covered with lavender flowers just as the foliage appears. Zone 10

86 *Gloriosa rothschildiana*

87 *Graptophyllum pictum*

Gloriosa rothschildiana — Climbing Lily / Rothschild Glorylily

Family: Liliaceae
Origin: South Africa

86 A tuberous, climbing herb to 8 ft. The 8 in. wide, ribbon-like, recurved segments are yellow to red before becoming a dull red with age. Both the tubers and plants are highly toxic if eaten. Zone 9

Graptophyllum pictum (G. hortense) — Caricature Plant

Family: Acanthaceae
Origin: New Guinea

87 Widely grown for its variegated foliage, this 6–8 ft. evergreen plant has 4–6 in. leaves that are either plain green or reddish-purple and oval-pointed at both ends. Small clusters of crimson-purple, tubular, ½ in. long flowers are produced. Zone 10

88 Grevillea 'Noell'

Grevillea 'Noell'

Family: Proteaceae
Origin: Australia

88 A clean, low, compact evergreen plant with needle-like, bright green foliage and rose-red blooms in the spring. It is an excellent sunny bank cover, which can be kept to 3 ft.   Zone 8

Hedychium coronarium   Garland Flower
                                          White Ginger

Family: Zingiberaceae
Origin: India; Indonesia

89 A perennial growing to about 6 ft., with 6 in. long, fragrant flower clusters of pure white. In India, it is considered the most attractive of all gingers.   Zone 9

89 Hedychium coronarium

Hedychium gardnerianum   Kahili Ginger

Family: Zingiberaceae
Origin: India

90 Another species from the extensive genus of ginger one sees in the tropics and subtropics. This perennial grows to about 6 ft. The flower is light yellow, with red stamens. Summer-flowering.   Zone 9

90 Hedychium gardnerianum

HELICONIA

Family: Heliconiaceae
91–100 The large, colorful bracts, which hide the tiny flowers, often are up to 6 in. long and 2 in. wide.   Zone 10 and Above

91   *Heliconia bihai* var. *aurea*

91A   *Heliconia bihai* var. *aurea*

92   *Heliconia caribaea*

Heliconia bihai var. *aurea (H. distans)*   Firebird
Macaw Flower
Wild Plantain

Origin: Northern South America
91   Similar to *H. caribaea* but shorter, with reddish-orange bracts fading to the yellow edges. Grows to 6 ft. with up to 5 ft. leaves.

Heliconia caribaea

Origin: West Indies
92   A tall-growing variety with 8 ft. stalks and 3–5 ft. long leaves, a foot wide. The flowers are yellowish with brownish bracts.

93   *Heliconia humilis*

93A   *Heliconia humilis*

96   *Heliconia lingulala*

94   *Heliconia lataspathia*

95   *Heliconia esperito santos*

97   *Heliconia pendula*

Heliconia humilis   Lobster Claw

Origin: Northern South America

93   The stalks grow to 4 ft. with leaves up to 5 ft. long. The bracts are dark red with a small, green edge.

Heliconia lataspathia

Origin: Central America

94   The stalks are to 10 ft., with 3–5 ft. leaves. The inflorescences are upright, usually borne above the leaves, and are 12–20 in. high. The bracts are arranged spirally and are yellowish-green through bright orange.

Heliconia esperito santos

95   *A little-known variety from Brazil.*

Heliconia lingulala

Origin: Unknown

96   One of the few pure yellow varieties.

Heliconia pendula (H. collinsiana)

Origin: Central America

97   The stalks are up to 6 ft., with 3 ft. leaves, which are a foot wide. The flowerhead is up to 18 in. long.

8   *Heliconia psittacorum*

98A   *Heliconia psittacorum*

99   *Heliconia rostrata*

100   *Heliconia wagnerana*

Heliconia psittacorum   **Parrot's Flower**
Origin: Northern South America   **Parrot's Plantain**
98   This dwarf form is shown growing to about 3 ft. in clumps, usually in or around a lawn. It has many orange bracts, which last for long periods of time.

Heliconia rostrata
Origin: From Peru to Argentina
99   The stalks are to 5 to 6 ft., with 4 ft. leaves. The pendulous bracts, up to 15 in., are red at the base with yellow tips.

Heliconia wagnerana (H. elongata; H. stricta)
Origin: Costa Rica
100   Very exotic looking, with a pale crimson color and cream at the base with a green edge. Stalks to 8 ft. with 3–5 ft. leaves.

Hemerocallis lilioasphodelus (H. flava)   **Day Lily**
Family: Liliaceae   **Lemon Lily**
Origin: Eastern Siberia to Japan
101   The popular Day Lily and its many cultivars are widely grown from tropical Hawaii to those regions that sustain temperatures below zero during the winter. The flowers close at night, never to reopen, but there is a plentiful supply to provide flowers for many days.   Zone 5
   Picture from Lyons Botanical Garden in Honolulu.

Hemigraphis alternata (H. colorata)   **Red Flame**
Family: Acanthacea   **Red Ivy**
Origin: Java
102   A creeping, perennial ground cover, with purplish leaves showing some green above but purple underneath. Grown as a pot plant in colder areas. It is used medicinally to treat skin diseases.   Zone 10

101   *Hemerocallis lilioasphodelus*

102   *Hemigraphis alternata*

HIBISCUS

103   *Hibiscus moscheutos*

104   *Hibiscus rosa-sinensis* 'White Wings'

104A   *Hibiscus rosa-sinensis* 'Crown of Bohemia'

104B   *Hibiscus rosa-sinensis* 'San Diego Red'

Hibiscus moscheutos   Mallow Rose
                                      Swamp Rose Mallow
                                            Wild Cotton

Family: Malvaceae
Origin: Eastern North America
103   Many cultivars of this species are available. The plants are deciduous, 8 ft. tall, with large, single flowers, often up to 8 in. across, in red, pink, or white.   Zone 7

Hibiscus rosa-sinensis Cultivars
Family: Malvaceae
Origin: Garden Hybrids
104–104BB   These lovely plants bloom the year around. The flowers of the cultivars are 4–6 in. across and produced in many colors—white, yellow, pink, red—both single and double and grow from 5–15 ft. tall. They are grown in all tropical and semitropical regions of the world.   Zone 10

104C

104D

104E

104F

104G

104H

104I

104J

104K

Hibiscus Assorted Cultivars

Family: Malvaceae
Origin: Garden Hybrids
104C–104K   These are some of the latest cultivars introduced into the trade. The name given to each may vary from one geographic region to another.

105   *Hibiscus schizopetalus*   106   *Hibiscus tiliaceus*

Hibiscus schizopetalus   Japanese Hibiscus
                                            Japanese Lantern

Family: Malvaceae
Origin: Tropical East Africa
105   A graceful shrub to about 10 ft., with dainty, orange-red
flowers. It is one of the principal parents of the many beautiful
hibiscus cultivars we have today.   Zone 10

Hibiscus tiliaceus (H. abutiloides;   Hau Tree
Paritium liliaceum)   Mahoe Tree

Family: Malvaceae
Origin: Tropical America
106   This is the evergreen bush or tree hibiscus, 10–20 ft., often
used for arbors. The flowers last for only a single day. They are
bright yellow in the morning, and, as the day progresses, change to
apricot, then to red at the day's end. The bark is used for caulking
boats.   Zone 10

107   *Hippeastrum vittatum*

Hippeastrum vittatum (Amaryllis   Amaryllis
vittatum)   Barbados Lily

Family: Amaryllidaceae
Origin: Peru
107   A herbaceous, bulbous plant with thick flower stalks to 2 ft.,
each bearing several, large, trumpet-shaped flowers. The colors
are many, including red, orange, white, and variegated. The bulb is
poisonous and may cause a skin rash. Spring/early summer-
flowering.   Zone 9

Holmskioldia sanguinea   Chinese Hat Plant
                                            Mandarin's Hat
Family: Verbenaceae
Origin: Himalayas
108   An evergreen shrub that grows to 30 ft. The brick-red to
orange, and occasionally yellow, flowers are very unusual in that
they resemble a Chinese coolie's hat. It is common throughout the
tropics.   Zone 10

108   *Holmskioldia sanguinea*

Homalocladium platycladum   Centipede Plant
(Muehlenbeckia platyclada)   Tapeworm Bush
Family: Polygonaceae
Origin: Solomon Islands
109   This odd-looking plant grows to about 12 ft., with interesting, flattened, many-jointed stems and tiny, stalkless, white
flowers. The small, purple fruit appears at the joints. Drought resistant.   Zone 10

109   *Homalocladium platycladum*

110   Hylocereus undatus

111   Ixora coccinea

111A   Ixora coccinea

112   Ixora duffii

112A   Ixora duffii

113   Ixora odorata

113A   Ixora odorata

Hylocereus undatus   Night-blooming
                                                              Cereus
Family: Cactaceae   Queen of the Night
Origin: Mexico to northern South America

110   A fragrant, night-flowering cactus widespread throughout the tropics. The green, triangular, climbing stems of this epiphytic plant produce 10–12 in. long, white flowers as night falls, which wither away as the sun rises the next morning. Fruit red, edible.   Zone 10

Ixora coccinea (I. bandhuca; I. incarnata; I. lutea)   Burning Love
                                                              Jungle Flame
Family: Rubiaceae   Jungle Geranium
Origin: India

111   One of the most colorful of all tropical plants, it is grown both as a hedge and as a 4 ft. foundation planting. The clusters of coral-red flowers, about 3 in. across, create a vivid mass on the bush, covering it almost daily throughout the year. They make excellent cut flowers. The plant needs moist soil and will not stand frost. In colder areas it is grown as an indoor pot plant.   Zone 10

Ixora duffii (I. macrothyrsa)   Giant Ixora
                                                              Malay Ixora
Family: Rubiaceae
Origin: Caroline Islands; Sumatra

112   This shrub is similar to I. coccinea but grows to 10 ft. or more, with leaves and flowers twice the size. Its 5 in. flower clusters are scarlet and bloom throughout the year. It is one of the showiest plants in the tropics.   Zone 10

Ixora odorata   Ixora
Family: Rubiaceae
Origin: Madagascar

113   The flowers of this 10 ft. shrub grow in clusters, often 1 ft. across. They are very fragrant, with purple stems and a tube that is 4–5 in. long.   Zone 10

114   Jatropha integerrima

Jatropha integerrima (J. hastata;   Peregrina
Adenoropium hastatum;
A. integerrimum)

Family: Euphorbiaceae
Origin: Cuba

114   An evergreen shrub to about 10 ft. It grows in the sun, with
leaves up to 5 in. long. The clusters of bright red, 1 in. flowers
appear the year around. They develop into speckled, toxic seeds,
which scatter as the dry fruit explodes.   Zone 10

Jatropha multifida (Adenoropium   Coral Plant
multifidium)   Physic Nut

Family: Euphorbiaceae
Origin: Tropical America

115   A shrub or tree to 20 ft. with deeply divided leaves about 1 ft.
across. The flowers are a rich scarlet in July.   Zone 10

Justicia betonica (Nicoteba betonica)   Squirrels Tail
                                                   White Shrimp Plant

Family: Acanthaceae
Origin: Malaysia West to Tropical Africa

116   A weak-stemmed, 5 ft. shrub. The attractive inflorescences
are borne on erect spikes about 4 in. long. White bracts hide the
tiny, lilac-colored flowers.   Zone 10

114A   Jatropha integerrima

115   Jatropha multifida

116   Justicia betonica

117 *Justicia brandegeana*

118 *Justicia carnea*

Justicia brandegeana (Beloperone guttata; Drejella guttata) False Hop Shrimp Plant

Family: Acanthaceae
Origin: Mexico

117 An evergreen shrub to 3–4 ft., sometimes larger. Tubular, drooping, purple flowers spotted with white are enclosed in coppery bronze bracts which overlap to form the 3–4 in. long inflorescence. These somewhat resemble a large shrimp. The plant does best in partial shade and should be pruned regularly to shape. Zone 10
Also shown: *J. brandegeana* 'Yellow Queen,' a yellow cultivar of the above species.

Justicia carnea (Jacobinia carnea) Brazilian Plume Flower

Family: Acanthaceae
Origin: Brazil

118 This evergreen shrub to 6 ft. is best grown in rich, shaded soil. It is common in tropical areas, but, in colder regions of the world, is an old favorite in conservatories. Its flowers grow in clusters, pink to reddish in color. Zone 10

Lantana camara Common Lantana Shrub Verbena

Family: Verbenaceae
Origin: Southern North America to northern South America

119 Flowers of yellow to pink to orange or red grow in clusters on upright stems on this 10 ft. evergreen plant. It is widespread throughout the tropics. At one time it threatened agriculture with its rampant growth; however, parasites were introduced, slowing down its spread. The entire plant is toxic to cattle. Zone 10
Also shown: *L. c.* 'Radiation' and *L. c.* 'Yellow' (cultivars).

119 *Lantana camara*

119A *Lantana camara*

120   *Lantana montevidensis*

Lantana montevidensis (L. delicata;   Trailing Lantana
L. delicatissima; L. sellowiana)   Weeping Lantana

Family: Verbenaceae
Origin: South America

120   This trailing, evergreen shrub has been used widely in breeding the cultivars sold in nurseries today. It does best in full sun, withstands much drought, and blooms the year around. The entire plant, especially the berries, is toxic to cattle and can cause a skin irritation in humans.   Zone 10

Leucadendron argenteum   Silver Tree

Family: Proteaceae
Origin: South Africa

121   A large, evergreen shrub or tree to 20 ft., with blue-gray, silvery leaves. It needs full sun and very fast-draining soil.   Zone 10

Leucophyllum frutescens (L. texanum)   Barometer Bush
                                                 Texas Ranger
                                                 Texas Silver Leaf

Family: Scrophulariaceae
Origin: Southwestern Texas and Mexico

122   An evergreen shrub to 10 ft., it is slow-growing, takes any degree of heat (even desert), and thrives with very little water. The foliage is silver-colored. The shrub is used in landscaping, either as a roundheaded mass or as a clipped hedge.   Zone 8

121   *Leucadendron argenteum*

122   *Leucophyllum frutescens*

123   *Leucospermum reflexum*

123A   *Leucospermum cordifolium*

Leucospermum   **Pincushion Flower**

Family: Proteaceae
Origin: South Africa

123–123C   This is a genus of more than 40 species from which
many cultivars have been derived, all native to South Africa, and
the most spectacular plants grown there. They range in size from
2–10 ft. The plants need a well-drained soil, sun, and very little
summer water. The flowers are shipped all over the world as a
long-lasting cut flower.   Zone 9

123B   *Leucospermum cordifolium*

123C   *Leucospermum reflexum*

124   *Malvaviscus arboreus*

125   *Medinilla magnifica*

124A   *Malvaviscus arboreus*

Malvaviscus arboreus   Sleeping Hibiscus
                                                        Wax Mallow

Family: Malvaceae
Origin: Mexico to Brazil
124   A shrub to 10 ft., which flowers the year around. The evergreen leaves are very similar to those of the hibiscus, and the bright
red flowers look very much like those of the red hibiscus about to
open.   Zone 10

Medinilla magnifica   Malaysian Orchid

Family: Melastomataceae
Origin: Philippines
125   An epiphyte, this choice, ornamental, evergreen shrub to 8
ft. has pendulous, pink panicles and 1 in. wide flowers of pink to
coral-red.   Zone 10
      Picture taken at Hilo, Hawaii Airport.

125A   *Medinilla magnifica*

126   *Megaskepasma erythrochlamys*

128   *Murraya paniculata*

129   *Musa × paradisiaca*

127   *Microsorium scolopendria*

Megaskepasma erythrochlamys   Brazilian Red Cloak
Family: Acanthaceae
Origin: Brazil
126   This evergreen shrub grows to about 6 ft. The leaves are up
to 1 ft. long. Many conspicuous, brilliant red, 1 ft. long spikes of
bracts hide tiny, white flowers.   Zone 10

Microsorium scolopendria (Polypodium   Laua'e
phymatodes; P. scolopendria)   Maile-scented Fern
Family: Polypodiaceae
Origin: Polynesia
127   This fern is found both wild and cultivated, creeping into
garden areas and up tree trunks, and is used extensively as a
ground cover in Hawaii and elsewhere, especially Polynesia. It has
broad, flat, oblong, shiny, dark green fronds, 1–3 ft. long.
                                                        Zone 10

Murraya paniculata (M. exotica)   Chinese Box
                                          Orange Jessamine
Family: Rutaceae   Satinwood
Origin: Southeast Asia; Malay Peninsula
128   A shrub to 10 ft., its fragrant, white flowers bloom several
times a year. It is widely used in the tropics both as a hedge and as a
specimen plant.   Zone 10

Musa × paradisiaca (M. acuminata ×   Edible Banana
balbisiana; M. × sapientum)   Plantain
Family: Musaceae
Origin: Garden origin
129   When Hawaii was discovered by the Europeans, the
Hawaiians had selected more than 50 cultivars of this cross and
since then who knows how many more have been used. The plants
are evergreen, monocarpic (they die after the flowers/fruits are
produced), and grow to a height of 15 ft.   Zone 10

130A   *Mussaenda erythrophylla*

131   *Mussaenda philippica* 'Donna Aurora'

130   *Mussaenda erythrophylla*

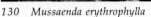

Mussaenda erythrophylla   Ashanti blood
Family: Rubiaceae   Red Flag Bush
Origin: Tropical West Africa

130   A very colorful, evergreen shrub to a height of 30 ft., with 6 in. long, oblong leaves. At times, the foliage is almost covered with large bracts that hide (or overpower) the tiny, sulfur-yellow or white flowers. The bract colors vary from light pink to coral to rose-red according to variety.   Zone 10

Mussaenda philippica 'Donna Aurora'   Buddha's Lamp
Family: Rubiaceae   Virgin Tree
Origin: Philippines

131   This showy member of the coffee plant family has white flower bracts covering tiny, yellow flowers. The plant was named for the wife of a Philippine president.   Zone 10

Nandina domestica   Heavenly Bamboo
                                                             Sacred Bamboo
Family: Berberidaceae
Origin: India to Japan

132   A very useful, evergreen shrub that thrives in both sunny and shady locations. It needs water during the summer. In colder areas the leaves turn a brilliant red in the fall. The plant reaches a height of 8 ft.   Zone 6

132   *Nandina domestica*

Nerium oleander   Oleander
(N. indicum; N. odorum)   Rose Bay
Family: Apocynaceae
Origin: Mediterranean

133   An evergreen shrub to 20 ft., commonly planted in the tropics. The single, sometimes double, flowers are red, pink, rose, or white. All parts of the plant are poisonous to man, animals, and insects when eaten. It grows best in full sun, is drought resistant and salt tolerant.   Zone 9

Also shown: 'Petite Pink,' a new dwarf form. It has the same characteristics but grows only about 4 ft. in 10 years.

133   *Nerium oleander*

133A   *Nerium oleander*

134 *Nicolaia elatior*

135 *Norantea guianensis*

135A *Norantea guianensis*

136 *Ochna kirkii*

137 *Odontonema strictum*

Nicolaia elatior (Phaeomeria magnifica; P. speciosa) Philippine Wax Flower / Torch Ginger

Family: Zingiberaceae
Origin: Java

134 Grows in large clumps to 10–20 ft. in height with leaves that are 2 ft. long and 6 in. wide. The flowerhead grows on a 1 in. stalk from the ground and is made up of waxy, red bracts covering the tiny flowers. Zone 10

Norantea guianensis

Family: Marcgraviaceae
Origin: Guyana

135 A clean, shiny, climbing shrub, with long, narrow, reddish-orange inflorescence made up of many small flowers and their bracts. With support, it will grow to 10 ft. Zone 10
 The picture is from the Foster Gardens, Honolulu.

Ochna kirkii Bird's Eye Bush / Mickey Mouse Plant

Family: Ochnaceae
Origin: Southeastern Africa

136 An evergreen shrub to about 12 ft., which thrives only in warm areas and will stand only light frost. This interesting plant grows in light shade. In spring, after the new foliage appears, masses of bright yellow flowers, similar to the buttercup, cover the plant. Two months later the sepals and receptacle enlarge and turn scarlet, revealing a green seed, which turns black as it ripens. The name "Mickey Mouse" comes from the face of the fruit. Zone 10

Odontonema strictum Cardinal's Guard

Family: Acanthaceae
Origin: Central America

137 A shrub to 6 ft. with 3–6 in. long, narrow, pointed leaves and bright orange-red flower clusters on the tip ends of the branches. Best grown in partial shade. Zone 10
 Picture: Foster Gardens, Honolulu.

ORCHIDS

138–138R   Orchids belong to the Orchidaceae family. When one thinks of tropical plants, those of the orchid family usually are thought of first. It generally is considered to be the largest family in the plant kingdom, with more than 600 genera and 20,000–30,000 species. This book identifies only a few of the common genera. The following illustrations are typical of some of the orchids one might see on a tropical visit.

138A   *Dendrobium* hybrid

138   *Dendrobium* 'Thomas Warne'

138B   *Dendrobium* 'Portia'

138C Dendrobium nobile

138D Spathoglottis plicata

138E Oncidium spathulatum

138F Cattleya

138G Cattleya

138H Cattleya

138I Cattleya

138J Vanda

138K   *Phalaenopsis*

138L   *Phalaenopsis*

138M   *Vanda terete* 'Miss Joaquin'

138N   *Cymbidium*

138O   *Angraecum palcatum*

138P   *Epidendrum agvense*

138Q   *Epidendrum* hybrid

138R   *Epidendrum* hybrid

39 *Pachystachys lutea*

140 *Pandanus baptistii aureus*

achystachys lutea Golden Candle

amily: Acanthaceae
rigin: Peru

39 This plant grows in tropical areas to about 5 ft. and as broad.
early summer it is covered with golden "candles," which last for
veral months. It is grown outdoors in frost-free areas and as a
reenhouse pot plant in other areas. Zone 10

andanus baptistii aureus Timor Screw Pine

amily: Pandanaceae
rigin: East Indies; Timor

40 The leaves of this species grow to about 3 ft. long, with alter-
ating green and yellow stripes but without spines on the margins.
prefers a dry soil and full sun. Grows to 20 ft. Zone 10

141 *Pandanus odoratissimus*

andanus odoratissimus Pandang Screw Pine

amily: Pandanaceae
rigin: Sri Lanka to the Philippines

41 This species grows to 20 ft. The 3–6 ft. long leaves carry very
harp spines. As the plant grows older, the leaves drop from the
wer trunk, leaving foliage only at the top. Aerial roots that grow
the ground then appear to give this 20 ft., top-heavy tree needed
upport. The leaves are used for weaving baskets, mats, and
hatching. Zone 10

ennisetum setaceum (P. ruppelianum; Fountain Grass
. ruppelii)

amily: Graminea
rigin: Africa

42 An attractive, large, coarse, perennial grass often grown in
ardens and frequently found in the wild as a garden escape. It
rows in bushy tufts, 2–3 ft. tall, with feathery, pink or purple
lowering spikes about 1 ft. taller, and is gray-green in
olor. Zone 9

142 *Pennisetum setaceum*

143   Philodendron evansii

Philodendron evansii

Family: Araceae
Origin: Garden Hybrid

143   One of the many cultivars between *P. selloum* × *P. speciosum* found in gardens in many parts of the tropical and semitropical world. The leaves often are more than 2 ft. wide and 3–4 ft. long. The plant is 10 ft. high and as wide.   Zone 10

144   Philodendron selloum

Philodendron selloum (P. johnsii)

Family: Araceae
Origin: Brazil

144   The hardiest of the big-leafed philodendrons, it is used out of-doors in areas to about 25°. It has deeply cut leaves about 3 ft long, and a single plant can grow to 10 ft. and as wide.   Zone 10

145   Plumbago auriculata

Plumbago auriculata (P. capensis)   Cape Plumbago
                                                                                             Leadwort

Family: Plumbaginaceae
Origin: South Africa

145   If unsupported, this plant is a mounding shrub to 6 ft., but with support, it will reach 12 ft. or more. The light-blue, sometimes white, 1 in. flowers are borne in clusters and bloom from March to December. The plant must have good drainage. All parts are toxic.   Zone 9
    Photo taken at the Honolulu Airport.

PROTEA

Family: Proteaceae
Origin: South Africa
146–146G   The King Protea, shown here, is one of many species
in the Family Proteaceae, which are all commonly called protea.
They are native to South Africa and are used in international cut-
flower markets and as garden plants. The soil must be free of lime,
ashes, or manure. Other genera in this family include *Banksia,*
Leucodendron, and *Leucospermum.* Although *Protea* is a genus native
to South Africa, other members in this family come from Australia.
All are evergreen.   Zone 9

46   *Protea eximia*

146A   *Protea scolymocephala*

46B   *Protea neriifolia*

146C   *Protea neriifolia*

146D   *Protea magnifica*

146E   *Protea sulphur*

146F   *Protea cynaroides*

146G   *Protea neriifolia*

47 *Pseuderanthemum reticulatum*

148 *Psidium cattleianum*

149 *Reinwardtia indica*

150 *Rhoeo spathacea*

seuderanthemum reticulatum   Eldorado
Eranthemum reticulatum)

amily: Acanthaceae
rigin: Probably southern Polynesia
47   This handsome, dark green and yellow, variegated hedge
lant will grow close to the sea. It usually is kept trimmed to 3–4
                                                    Zone 10

sidium cattleianum (P. littorale)   Strawberry Guava

amily: Myrtaceae
rigin: Brazil
48   An evergreen shrub or small tree to 20 ft. with long, leathery,
ark green foliage. The 3–4 in., white flowers are followed by dark
ed fruit, 2 in. in diameter, which has a spicy flavor and is eaten
resh or preserved.   Zone 10

einwardtia indica (R. tetragyma;   Yellow Flax
. trigynum; Linum trigynum)

amily: Linaceae
rigin: Northern India; China
49   A showy, low-growing, evergreen shrub with showy,
rilliant yellow, 1–2 in. flowers and small, narrow leaves. It grows
 3 ft. and blooms in winter and spring.   Zone 10

hoeo spathacea (R. discolor;   Moses-in-a-Boat
. tradescantia discolor)   Moses-in-a-Cradle

amily: Commelinaceae
rigin: West Indies; Mexico
50   A perennial, succulent-type species rarely more than 1 ft.
ll. The accompanying illustration shows it used as a ground cover
n a highway in the tropics. In other parts of the world, it is a
opular house plant. The sap is toxic.   Zone 10

Russelia equisetiformis (R. juncea)   Coral Plant
                                                  Fire Cracker Plant
                                                    Fountain Bush

Family: Scrophulariaceae
Origin: Mexico
151   A drooping shrub to 4 ft., usually leafless, with slim, wispy
branches and 1 in. long, with bright red flowers in abundance the
year around. It wants partial shade with little water.   Zone 10

152   *Saccharum officinarum*

Saccharum officinarum   Sugarcane

Family: Gramineae
Origin: Tropical Southeast Asia
152   This grass is cultivated in most of the tropics and is the
source of cane sugar. It has 2 ft. long, narrow leaves, which are
removed (usually by burning) before the stems are harvested and
made into sugar. The overall height of the plant is 15 ft.   Zone 10

151   *Russelia equisetiformis*

153   *Sanchezia nobilis* var. *glaucophylla*

Sanchezia nobilis var. *glaucophylla*   Sanchezia

Family: Acanthaceae
Origin: Ecuador
153   A shrub to 6 ft. with large, evergreen leaves, which have
variegated mid- and side veins and are pointed at both ends. The
yellow flowers grow upright on 2 in. long spikes, with 1 in. long,
red bracts. It is a nice-looking plant but can be invasive and take
over a good part of a garden.   Zone 10

153A   *Sanchezia nobilis* var. *glaucophylla*

154 *Scaevola sericia*

Scaevola sericia (S. frutescens; S. taccada)   Beach Naupaka
Family: Goodeniaceae
Origin: Tropical Asia
154   A spreading, succulent shrub, 3–10 ft. It grows on the beaches and is used as a hedge or an ornamental, as well as a soil binder. The white flowers grow in clusters of five or more and appear to be split in two, with only a half flower remaining, followed by ½ in., white berries.   Zone 10

154A   *Scaevola sericia*

155   *Spathiphyllum* 'Clevelandii'

Spathiphyllum 'Clevelandii'   Mauna Loa
Family: Araceae   Spathe Flower
Origin: Central America; Malaysia
155   This evergreen perennial is grown throughout the tropics in shady areas. There are more than 30 species, which grow to 1½–4 ft. The flowers are white and resemble anthuriums. It is grown as a popular house plant elsewhere.   Zone 10

Stemmadenia galeottiana   Crape Jasmine
Family: Apocynaceae   Lecheso
Origin: Central America
156   This small, evergreen shrub grows to 10–12 ft., with a picturesque, crooked growth habit. The fragrant flowers are white with a yellow throat, funnel-shaped, and about 2 in. long.   Zone 10

156   *Stemmadenia galeottiana*

157   Strelitzia nicolai

158   Strelitzia reginae

158A   Strelitzia reginae

159   Strelitzia reginae var. juncea

Strelitzia nicolai   Giant Bird of
                                                       Paradise

Family: Strelitziaceae
Origin: South Africa

157   This palm-like, evergreen "Bird of Paradise," with its giant
leaves, grows to 20 ft. The flowers are blue and white. The plant
branches from the base. Each clump develops a dozen or more
thick trunks.   Zone 10

Strelitzia reginae (S. parvifolia)   Bird of Paradise
                                                       Crane Flower

Family: Strelitziaceae
Origin: South Africa

158   Widely grown in the tropics and subtropics, this exotic plant
is clump-forming and grows to 3–4 ft. The "Birds" are the orange or
yellow flowers with a dark blue tongue. They are long-lasting and
are shipped to florists the world over.   Zone 10

Strelitzia reginae var. *juncea (S. juncea;*   Small-leafed
S. parvifolia var. *juncea)*   Strelitzia

Family: Strelitziaceae
Origin: South Africa

159   The flowers of this plant are similar to those of *S. reginae* but
the leaves differ. They are very small and blade-like and appear at
the tips of stiff, rounded stalks. When the plant is mature, these
leaf-blades disappear, and a pointed stem without leaves is
formed.   Zone 10

160 *Streptosolen jamesonii*

160A *Streptosolen jamesonii*

Streptosolen jamesonii   Firebush Bush
Marmalade Bush

Family: Solanaceae
Origin: Colombia and Ecuador
160   A rambling, evergreen shrub to 6 ft. with dense clusters of 1 in., orange-red flowers at the end of each branch. It likes plenty of sun.   Zone 10

Tapeinochilus ananassae   Giant Spiral Ginger

Family: Zingiberaceae
Origin: East Indies
161   A very interesting plant growing to 7 ft., with bamboo-like canes, it is a member of the ginger family. The "flower" resembles a pineapple in size but actually consists of orange-red bracts nesting small, yellow flowers.   Zone 10

Tecoma stans (Bignonia stans;   Yellow Elder
Stenolobium stans)

Family: Bignoniaceae
Origin: West Indies; northern South America; now naturalized in Florida
162   A deciduous shrub or small tree to 25 ft. It is beautiful for a short period in the fall, with bell-like, 2 in., yellow flowers that grow in pendulous clusters.   Zone 10

161 *Tapeinochilus ananassae*

162 *Tecoma stans*

163   *Tetrapanax papyriferus*

164   *Thevetia peruviana*

Tetrapanax papyriferus (Aralia   Rice-paper Plant
papyriferus; Fatsia papyrifera)
Family: Araliaceae
Origin: China; Taiwan
163   The white pith of this multi-stemmed, evergreen plant is the
source of rice paper in China. The large leaves, over 1 ft. across, are
gray-green and reach up to 8–20 ft. in height.   Zone 10

Thevetia peruviana (T. neriifolia)   Be-still Tree
Family: Apocynaceae   Lucky Nut
Origin: Tropical America   Yellow Oleander
164   This evergreen shrub grows to 30 ft. so should be pruned
heavily in windy areas. It will take any amount of heat and sun, but
will tolerate very little frost. It often is called the Yellow Oleander
but is not a member of the Oleander family but rather of the
Periwinkle family. The shrub is poisonous.   Zone 10

164A   *Thevetia peruviana*

Thunbergia erecta (Meyenia erecta)   Bush Clockvine
Family: Acanthaceae   King's Mantle
Origin: Tropical Africa
165   A bushy, sprawling, evergreen shrub to 8 ft. The flowers are
single, bell-shaped, about 3 in. long, and violet with a yellow throat
in July.   Zone 10

165   *Thunbergia erecta*

165A   *Thunbergia erecta*

166 *Tibouchina urvilleana*

167 *Tulbaghia violacea*

168 *Victoria amazonica*

169 *Warszewiczia coccinea*

169A *Warszewiczia coccinea*

Tibouchina urvilleana (T. grandiflora;	Glory Bush
T. semidecandra; Pleroma grandiflora;	Princess Flower
P. splendens)	

Family: Melastomataceae
Origin: Brazil

166 Found only in a few tropical plantings, this evergreen plant has gone wild around Volcano, Hawaii. It has 2–3 in., purple flowers growing in profusion during the summer, making the ground seem as if it is covered by a purple carpet. It grows rapidly to 15 ft. and is somewhat leggy, but will withstand heavy pruning.   Zone 10

Tulbaghia violacea   Society Garlic

Family: Amaryllidaceae
Origin: South Africa

167 The onion-like foliage of this bulbous plant is about 1 ft. tall. Bright lilac-pink flowers grow on the stems up to 2 ft.   Zone 10

Victoria amazonica (V. regia var. randi)	Amazon Water Lily
	Royal Water Lily
Family: Nymphaeaceae	Water-Platter
Origin: Guyana; Amazon Region	

168 A gigantic water lily named for Queen Victoria of England. The leaves, 3–6 ft. in diameter, lie flat on the water and are able to support the weight of a child. It is widely used in quiet ornamental ponds throughout the tropics.   Zone 10

Warszewiczia coccinea   Wild Poinsettia

Family: Rubiaceae
Origin: Central America to Brazil

169 The plant grows to 20 ft. or more, with brilliant, 2 ft. long, 1 ft. wide clusters of bright red bracts, which hide tiny, yellow flowers. The leaves are 2 ft. long and 1 ft. across. It will not tolerate frost.   Zone 10

170   Wedelia trilobata

171   Yucca aloifolia

171A   Yucca aloifolia

172   Yucca gloriosa

173   Yucca whipplei

Wedelia trilobata   Trailing Wedeli

Family: Compositae
Origin: Southern Florida to tropical Central America

170   A vigorous, dark green ground cover, it grows in sun o
shade. As it grows, the trailing stems, 6 ft. long or more, root and s
spread the plant. The flowers are bright yellow to orange. It i
tolerant of salt spray. The seeds are toxic.   Zone 1

Yucca aloifolia   Dagger Plan
                                                         Spanish Bayone
Family: Agavaceae
Origin: Southwestern United States;
northern Mexico

171   This plant grows to 20 ft., with a rigid trunk, often branche
with very sharp, pointed, dark green leaves, which are 1–2 in. wid
and 2–3 ft. long. The flowers are white, held in erect spikes up to
ft. tall. When it becomes too tall, it can be cut back.
       Also shown is Y. a. 'Variegata'.   Zone

Yucca gloriosa   Roman Candl
                                                          Soft-tip Yucc
Family: Agavaceae   Spanish Dagge
Origin: Southeastern United States from
North Carolina to Florida

172   Generally multi-trunked, this plant grows to 10 ft. The 2–3 f
leaves are soft-tipped, pointed, and dark gray-green. The flower
appear in late summer on spikes and are greenish-white t
reddish.   Zone

Yucca whipplei (Hesperoyucca whipplei)   Our Lord's Candl

Family: Agavaceae
Origin: Southern California Mountains; Baja California

173   This plant grows with a dense cluster of 1–2 ft., rigid, gray
green leaves, which are needle-pointed. It is stemless. Th
fragrant, creamy white flowers appear on stalks that are up to 10 f
tall. The plant is monocarpic (dies after flowering).   Zone

TREES

74   *Adansonia digitata*

Adansonia digitata   Baobab
                                  Dead-rat Tree
Family: Bombacaceae   Monkey-bread Tree
Origin: Tropical Africa

74   A deciduous tree, seldom more than 60 ft. tall but up to 100 ft. in the circumference of the trunk. The 5–9 in. long, oval, hanging fruit has given it the name of "Dead-rat Tree."   Zone 10

175   *Albizia julibrissin*

Albizia julibrissin   Mimosa Tree
                                         Pink Acacia
Family: Leguminosae   Silk Tree
Origin: Iran to China

175   A deciduous tree that grows rapidly to 30–40 ft. and spreads as wide in the shape of an umbrella. It does best where it has hot summers, but does not do well near coastal areas. The flowers are pink, produced in the summer (July–August), with the heads crowded on the upper end of the branches.   Zone 7

Aleurites moluccana   Candebury Tree
                                        Candlenut Tree
Family: Euphorbiaceae   Kukui
Origin: Probably from the Moluccas but naturalized throughout the tropical world

176   A common, evergreen tree to 50 ft. From a distance the leaves appear whitish, or even variegated, due to a whitish down that clings to them. The leaves are shaped "more or less" like a maple leaf. Each tree can produce 75–100 pounds of nuts which have an oil content of 50%.   Zone 10

76   *Aleurites moluccana*

176A   *Aleurites moluccana*

177   *Araucaria araucana*

177A   *Araucaria araucana*

178   *Araucaria bidwillii*

179   *Araucaria heterophylla*

Araucaria araucana (A. imbricata)   Chilean Pine
                                                   Monkey Puzzle Tree

Family: Araucariaceae
Origin: Chile

177   This coniferous, evergreen timber tree grows to 50 ft. or
more in its native land. The awl-shaped, sharply pointed leaves are
1–2 in. long.   Zone 7

Araucaria bidwillii   Bunya-Bunya

Family: Araucariaceae
Origin: Australia

178   A bold; symmetrical tree to 100 ft. or more, with down
curving limbs covered with sharp-pointed, awl-shaped, brigh
green leaves. It develops large cones up to 7 × 9 in., which weigl
up to 10 pounds.   Zone

Araucaria heterophylla (A. excelsa)   Norfolk Island Pin

Family: Araucariaceae
Origin: Norfolk Island

179   A very formal conifer first imported into Hawaii to be grow
as a source of masts for early sailing ships. One of the few conifer
that grows well and so planted in tropical areas. It can grow to
height of 200 ft.   Zone 1

80 Artocarpus altilis

180A Artocarpus altilis

81 Bauhinia blakeana

181A Bauhinia blakeana

82 Bauhinia cumingiana

182A Bauhinia cumingiana

Artocarpus altilis (A. incisus)   Breadfruit

Family: Moraceae
Origin: Malaysia

180   An evergreen tree to 60 ft. with broad, deeply cut, shiny, dark green leaves, with 6–9 lobes. The fruit is 6 in. or more across and is smooth and bright green until ripe, when it turns yellowish. The fruit is an important food item in its native habitat. This is a tree connected with the voyage of the "H.M.S. Bounty."   Zone 10

Bauhinia blakeana   Hong Kong Orchid Tree

Family: Leguminosae
Origin: China (Hort)

181   This evergreen tree was propagated from a single tree found in Canton, China. The flowers, shaped like an orchid, are much larger, up to 6 in., than other species. It grows to 40 ft., the leaves partially dropping as it flowers October–March. Does not set fruit.   Zone 9

Bauhinia cumingiana

Family: Leguminosae
Origin: Philippines

182   Usually grows as a shrub to 6 ft. with large, bright green leaves. The flowers are in clusters, the petals of which are about 1 in. long and dull yellow with red markings. Flower in spring and early summer.   Zone 10

183   *Bauhinia galpinii*

184A   *Bauhinia monandra*

185A   *Bauhinia variegata* 'Candida'

184   *Bauhinia monandra*

185   *Bauhinia variegata* 'Candida'

Bauhinia monandra   Butterfly Flower
Family: Leguminosae   Pink Orchid Tree
Origin: Burma   St. Thomas Tree
184   Native to Burma but naturalized in tropical America and
commonly planted in the American tropics. The leaves are shaped
like a butterfly and are split into equal parts halfway in. It grows to
20 ft., with showy pink flowers about 4 in. across in March or April.
The thick seed pods are about 9 in. long.   Zone 10

Bauhinia variegata 'Candida'   Buddhist Bauhinia
Family: Leguminosae   Mountain Ebony
Origin: China, India   Orchid Tree
185   A tree to 20 ft. or more that is extremely variable during its
blooming period, depending on soil and weather. Occasionally it
drops its leaves in mid-winter and will then bloom with or without
foliage in very early spring and into summer. Will withstand con-
siderable cold but likes heat and a well-drained soil. There are
many colors of *B. variegata*, but the most commonly grown is
'Candida,' with pure white flowers and greenish veins.   Zone 9

Bauhinia galpinii (B. punctata)   Red Bauhinia
Family: Leguminosae
Origin: Tropical Africa to Transvaal
183   This beautiful, sprawling tree is deciduous in cooler areas. It
has lovely, brick-red to orange flowers and is excellent as an
espalier. Summer-flowering.   Zone 10

86   Bixa orellana

186A   Bixa orellana

186B   Bixa orellana

Bixa orellana   Achiote
                                                                  Annatto
Family: Bixaceae   Lipstick Tree
Origin: Central America

186   An evergreen tree or shrub to 20 ft., with large, heart-shaped, green leaves and dainty, pink or white, 2 in. flowers, followed by clusters of bright red or yellow seed pods that eventually turn brown. This was a major source of the red and yellow body paint used by the natives of Central America. It is now used commercially as a food coloring.   Zone 10

87   Bombax ceiba

ombax ceiba (B. malabaricum;   Red Silk Cotton Tree
almalia malabarica)   Simal

amily: Bombacaceae
rigin: India, Malaysia, Australia

87   A large, deciduous tree to 120 ft. heavily buttressed with few many spines on the trunk and branches. The flowers, which ppear in January, are red and similar to other *Bombax* species. The eed pods are produced in April and May; the seeds contain a large uantity of kapok but of inferior quality.   Zone 10

187A   Bombax ceiba

Brachychiton acerifolius   Australian Flame
Family: Sterculiaceae   Tree
Origin: Australia   Flame Tree
188   A deciduous tree to 100 ft. in all but the warmest areas,
usually dropping its leaves before flowering in early summer.
Slow-growing, with long, loose racemes of bright scarlet flowers
and a distinctive swollen trunk.   Zone 10

188   *Brachychiton acerifolius*

190   *Brassaia actinophylla*

191   *Brownea capitella*

189   *Brachychiton populneus*

Brachychiton populneus (Sterculia   Bottle Tree
diversifolia)

Family: Sterculiaceae
Origin: Australia

189   An evergreen tree to 40 ft. with a 30 ft. spread and a very
heavy trunk. There are small clusters of white flowers in May and
June.   Zone 9

Brassaia actinophylla (Schefflera   Octopus Tree
actinophylla)   Queensland
Family: Araliaceae   Umbrella Tree
Origin: Australia; Java; New Guinea

190   An evergreen tree to 40 ft. The unusual red flowers stand out
above the foliage on long arms resembling an octopus. Widely
used as a landscape plant in tropical and subtropical areas. Elsewhere it is grown indoors in tubs, where, by cutting back, it is kept
multistemmed and bushy.   Zone 10

Brownea capitella   Brownea

Family: Leguminosae
Origin: Venezuela

191   A handsome, small, evergreen tree to 30 ft., with long,
narrow leaves and showy, 11 in. clusters of small, reddish-brown
flowers in summer.   Zone 10

191A   *Brownea capitella*

92 *Butea monosperma*

192A *Butea monosperma*

Butea monosperma (B. frondosa)   Flame-of-the-Forest
                                     Parrot Tree
Family: Leguminosae
Origin: India, Burma, Pakistan
192   A deciduous tree to 40 ft., with compound, 8 in. leaves and beautiful, orange-red flowers growing in masses from January to March, when the tree has few, if any, leaves. The source of "Bengal Gum."   Zone 10

93 *Callistemon viminalis*

Callistemon viminalis   Weeping Bottlebrush
amily: Myrtaceae
Origin: Australia
93   The long, arching, pendulous branches each carry a mass of re-red bottlebrushes. It grows, tree form, to about 20 ft. but in its native habitat will grow to 60 ft. and is evergreen.   Zone 9

193A *Callistemon viminalis*

194A   *Calodendrum capense*

Calodendrum capense   Cape Chestnut

Family: Rutaceae
Origin: South Africa

194   This beautiful, evergreen tree grows to 60 ft. It flowers in midsummer, with large, showy clusters of rosy pink flowers. It is usually evergreen but may lose its leaves in colder areas. It will stand only a very light frost.   Zone 10

Cassia excelsa   Cassia
                                               Crown of Gold
Family: Leguminosae
Origin: Eastern Brazil

195   A large shrub or tree to 30 ft., with bright yellow racemes, up to 1 ft. long. The flowers are fragrant and, like most cassias, in bloom most of the year, but the heaviest bloom is in late summer and autumn.   Zone 10

195   *Cassia excelsa*

196   *Cassia fistula*

Cassia fistula   Golden Showe
                                            Indian Laburnun

Family: Leguminosae
Origin: India

196   This deciduous tree, to 30 ft., develops 1–2 ft. long raceme of pale yellow flowers, resembling large bunches of grape Flowering lasts from March to August but is at its peak i June.   Zone 1

97   Cassia javanica

98A   Cassia javanica × C. fistula

Cassia javanica   Apple Blossom
Family: Leguminosae   Pink-and-White
Origin: Java and Sumatra   Shower Tree

197   This low tree is loaded with masses of flowers, sometime between March and September, usually starting in June, holding in color for a long period, then carpeting the ground with the petals.   Zone 10

198   Cassia javanica × C. fistula

Cassia javanica × *C. fistula*   Rainbow Shower
Family: Leguminosae   Tree
Origin: A Hybrid

198   A deciduous, hybrid tree resulting from pollenizing the blossom of the Pink-and-White Shower Tree with the Golden-Yellow Shower Tree. This beautiful tree blooms from March to early September. Official tree of the city of Honolulu.   Zone 10

199   Casuarina equisetifolia

Casuarina equisetifolia   Australian Pine
                                                     Beefwood
Family: Casuarinaceae   Horsetail Tree
Origin: Australia and the Pacific Islands

99   A common evergreen tree in the tropics growing to 70 ft. It is idely planted as a hedgerow, often 2 ft. apart and kept clipped, ut more often it is used as a windbreak. It is salt-tolerant. This tree echnically is leafless, its fine, hair-like needles being fine-jointed ranches, hence the name taken from "equisetem."

200   *Ceiba pentandra*

201   *Cerbera manghas*

201A   *Cerbera manghas*

Ceiba pentandra (C. casearia;   Kapok
Eriodendron anfractuosum)   Silk-Cotton Tree

Family: Bombacaceae
Origin: Tropical America; Tropical Asia

200   A deciduous tree to more than 100 ft., with great width and a huge trunk, which is thorny when young. The leaves drop and the flowers, either creamy white or rose, develop into a 6 in. long pod containing a silky cotton, like kapok, which once was used in the manufacture of life preservers and cushions.   Zone 10

Cerbera manghas

Family: Apocynaceae
Origin: Pacific Islands from Australia to Madagascar

201   A large, bushy, evergreen tree or shrub to 20 ft., sometimes planted in warm regions as an ornamental. The shiny leaves are large, up to 8 in. long, and broad. White and pink flowers grow in clusters. The fruit, 2 in. in diameter, appears in the spring.
                                                                  Zone 10

Chorisia speciosa   Floss Silk Tree

Family: Bombacaceae
Origin: Southern Brazil and Argentina

202   This tall, deciduous tree grows to 40 ft. In the fall and winter it usually is covered with vivid pink flowers, 2–3 in. across. It is variable in color, however, white or yellowish and pink to violet. The tree can be recognized by the hundreds of rose-like thorns growing out of the trunk.   Zone 9

202   *Chorisia speciosa*

202A   *Chorisia speciosa*

Chrysophyllum cainito Caimito
Star Apple
Family: Sapotaceae
Origin: Tropical America

203 An evergreen tree to 50 ft. with nice-looking, oblong, pointed, short, 6 in. leaves, shiny above and brownish below. The summer flowers are small and purple-white. The fruit looks like an apple, 2–4 in. across. It is edible and may be eaten fresh but it must be ripe. Zone 10

Clusia rosea Autograph Tree
Scottish Attorney
Family: Guttiferae
Origin: Bahamas and West Indies

204 A magnificent tree from 20–50 ft. with large, thick leaves up to 8 in. long. It will grow close to the ocean as it tolerates salt spray. People often "autograph" the leaves. The saying is, "Fools names and fools faces are always seen in public places." It has not proved false. Flowers pink to white, 2 in. across. Zone 10

Coccoloba uvifera Kino
Platterleaf
Sea Grape
Family: Polygonaceae
Origin: Southern Florida, West Indies, northern South America

205 A characteristic evergreen tree grown close to the beach and up to 25 ft. in height. The leaves are as much as 8 in. across, leathery, glossy, and veined red. The flowers are white and grow in dense racemes up to 10 in. in length. The fruit resembles grapes and is used frequently for making jelly. Completely salt tolerant. Zone 10

203 *Chrysophyllum cainito*

204 *Clusia rosea*

204A *Clusia rosea*

205 *Coccoloba uvifera*

206   *Cochlospermum vitifolium*

206A   *Cochlospermum vitifolium*

Cochlospermum vitifolium   Buttercup Tree
                                                          Wild Cotton
Family: Cochlospermaceae
Origin: Mexico to northern South America

206   A drought-resistant, deciduous tree to 40 ft. It has bright
yellow flowers from January to May, developing a velvety seed
pod containing "cotton." Both single- and double-flowering forms
are in cultivation.   Zone 10

Cordia boissiere   Anacahutta

Family: Boraginaceae
Origin: Southern New Mexico and Texas to Mexico

207   An evergreen tree or shrub to 25 ft. The 1½ in. diameter
white flowers have a yellow center and grow in terminal
clusters.   Zone 10

206B   *Cochlospermum vitifolium*

207   *Cordia boissiere*

207A   *Cordia boissiere*

208 *Cordia sebestena*

208A *Cordia sebestena*

Cordia sebestena — Geiger Tree

Family: Boraginaceae
Origin: Venezuela to the Florida Keys

208 An evergreen tree to 30 ft. that blooms orange to red and is in flower almost the year around. The flowers grow in clusters of buds, with four open at the same time, followed by small, white, nut-like, edible fruit. The dark green leaves are about 6 in. long and rough and stiff. It is widely cultivated in the tropics. John Audubon named the tree for John Geiger, who was a boat pilot in the Florida Keys during the 1830s. Zone 10

Couroupita guianensis — Cannonball Tree

Family: Lecythidaceae
Origin: Guiana

209 A huge, deciduous tree to 50 ft., planted as a novelty in many tropical gardens. The fragrant flowers, pink with a cream center, about 3 in. in diameter, push out of the bark on the main stem of the tree not attached to the foliage at the top as in most trees. They are followed by hard-shelled "cannonballs," 6–8 in. in diameter. The inner pulp of the fruit is ill-smelling. Zone 10

209 *Couroupita guianensis*

209A *Couroupita guianensis*

209B *Couroupita guianensis*

Cupaniopsis anacardioides (Cupania   Carrotwood
anacardioides)   Tuckeroo

Family: Sapindaceae
Origin: Australia

210   An ideal evergreen tree to 40 ft. for subtropical gardens. It is a clean tree, so is handsome as a patio, lawn, or street tree. Fragrant, greenish-white flowers turn to reddish seed pods. Withstands salt wind on the coast or hot, dry winds inland.   Zone 9

210   *Cupaniopsis anacardioides*

211   *Delonix regia*

211A   *Delonix regia*

211B   *Delonix regia*

Delonix regia (Poinciana regia)   Flamboyant
       Flame Tree

Family: Leguminosae   Royal Poinciana
Origin: Madagascar

211   One of the most beautiful trees in the world. It starts to bloom in early spring and flowers until late summer. The long seed pods remain for an outstanding length of time. It has beautiful, fern-like foliage, and grows to a height of 40 ft.   Zone 10

212   Dombeya wallichii

212A   Dombeya wallichii

Dombeya wallichii   African Mallow
                                                      Hydrangea Tree

Family: Sterculiaceae
Origin: Madagascar

212   A small tree to 30 ft. is sometimes grown as a large shrub.
The leaves are very broad and up to 12 in. long. The small, 1 in.
flowers, coral-pink to red, which start to bloom in August, form a
cluster, making a ball about 3 in. in diameter. They hang on for
some time after they turn brown.   Zone 10

Dracaena draco   Dragon Tree

Family: Agavaceae
Origin: Canary Islands

213   This odd-looking tree grows to about 40 ft. or more. It is
broad, with a thick, bare trunk and a crown of many stubby
branches, each bearing crowded, sword-like, 2 ft. long leaves. The
flowers are small, from which bright orange berries develop. It
grows to a great age. The sap, after being dried, is called Dragon's
Blood, and used both medicinally and for coloring varnishes used
on fine furniture.   Zone 9

Eriobotrya japonica   Japanese Medlar
                                                      Japanese Plum
                                                      Loquat

Family: Rosaceae
Origin: Eastern China and Japan

214   This drought-resistant, evergreen tree grows to 20 ft. in
height and as wide. It is widely cultivated, even in subtropical
areas, for its fruit as well as its ornamental value. The large, thick,
leathery leaves are very attractive.   Zone 7

213   Dracaena draco

214A   Eriobotrya japonica

214   Eriobotrya japonica

215   *Erythrina* × *bidwillii*

Erythrina × *bidwillii*   Bidwill Coral Tree
Family: Leguminosae
Origin: Garden Hybrid
215   This is a large, deciduous shrub rather than a tree, growing
to about 15 ft. It is a garden hybrid, a cross between *E. crista-galli*
and *E. herbacea.* There is an abundance of bright red flowers from
spring to winter. Should be cut back hard but beware of the many
thorns.   Zone 10

215A   *Erythrina* × *bidwillii*

216   *Erythrina caffra*

Erythrina caffra (E. constantiana)   Coral Tree
                                              Kaffirboom
Family: Leguminosae
Origin: South Africa
216   A large, broad, briefly deciduous tree up to 60 ft. high and
spreading. Clusters of 6 in. wide, scarlet flowers appear in early
spring, however, the colors may be variable. In March the flowers
give way to fresh, light green foliage. The tree may be pruned
severely to keep it within bounds, and it must be grown in a frost-
free area.   Zone 10

216A   *Erythrina caffra*

17 *Erythrina coralloides*

218 *Erythrina crista-galli*

18A *Erythrina crista-galli*

Erythrina coralloides   Naked Coral Tree

Family: Leguminosae
Origin: Central Mexico to Arizona

217 A deciduous, somewhat prickly tree to 20 ft., flowering before the leaves appear. The flowers are triangular, 3–4 in. long, brownish, and covered with hair. The seeds are red.   Zone 10

Erythrina crista-galli (E. laurifolia)   Cockspur Coral Tree
                                             Cry-Baby Tree
Family: Leguminosae
Origin: Brazil

218 A deciduous tree that grows to 10 ft. or more. It starts to bloom in early spring with long spikes of showy, pea-shaped, orange-red flowers. It continues to flower lightly until late fall. It is the hardiest of the *Erythrina* group and will do well in most of Zone 9.   Zone 9

Erythrina fusca (E. glauca)   Swamp Immortelle

Family: Leguminosae
Origin: Tropical Asia and Polynesia

219 This deciduous tree grows to 30 ft. or more in height. It differs from most other *Erythrina* in that the flower is beige.
                                             Zone 10

19 *Erythrina fusca*

219A *Erythrina fusca*

220   Erythrina humeana

Erythrina humeana   Natal Coral Tree

Family: Leguminosae
Origin: South Africa

220   This species has smaller flowers than most but they are very
showy. Each flower stem carries about a dozen bright red flowers,
away from the foliage. Blooms from August to December. Evergreen in warmer areas.   Zone 10

220A   Erythrina humeana

Erythrina variegata (E. indica)   Indian Coral Tree
                                        Tiger's Claw
Family: Leguminosae
Origin: Indonesia; Philippines

221   A deciduous tree widely used in the tropics and growing to
60 ft. and as broad. It flowers when leafless and is covered with
large, golden, variegated leaves for the rest of the year.   Zone 10

221   Erythrina variegata

221A   Erythrina variegata

222A   *Eucalyptus deglupta*

22   *Eucalyptus deglupta*

Eucalyptus deglupta   Mindanao Gum

Family: Myrtaceae
Origin: Philippines

222   This is one of the few *Eucalyptus* not native to Australia. It grows to well over 100 ft. tall. The bark is unusual and attractive.   Zone 10

Fagraea berteriana   Pua-kenikeni

Family: Loganiaceae
Origin: Australia and the Pacific Islands

223   A tree to 40 ft. but often grown as a shrub. It has bright, shiny leaves about 6 in. long and 3 in. wide. The flowers are fragrant and used for perfumes and in leis.   Zone 10

23   *Fagraea berteriana*

223A   *Fagraea berteriana*

Ficus auriculata (F. roxburghii) Roxburg Fig
Family: Moraceae
Origin: Himalayas

224 A tree or large shrub to 20 ft. The leaves are 1 ft. wide and about 18 in. long, rounded, heart-shaped, dark, and leathery. The fruit, often found at the base of the trunk, is sweet and edible.
Zone 10

Ficus benghalensis (F. indica) Banyan Tree
Family: Moraceae East India Fig
Origin: India

225 An evergreen tree, with oval, dark green, broad, 10 in. leaves that are downy underneath and shiny above. This is one of the world's largest trees, many with a spread of 200 ft. across; one in India is said to measure 2,000 ft. in circumference at the dripline. As the branches grow out and away from the trunk, they develop descending members. These descending members make contact with the ground root to feed, forming additional trunks which support the branch, thus preventing breakage during storms. The Hindus believe that Brahma was transformed into a Banyan Tree and hold the tree sacred. Zone 10

Ficus benjamina Small-leaf Rubber Tree
Family: Moraceae Weeping Fig
Origin: India; Malaysia

226 This popular house plant grows ''like a weed'' in warm, humid areas, which results in a tree about 50 ft. high and as wide. It is widely planted in the tropics. It should not be planted near sidewalks or streets, not only because of the damage done to them by the roots but also because of the hazard caused by the messy fruit. A popular house plant in colder areas. Zone 10

Ficus elastica (F. belgica) India Rubber Plant
Family: Moraceae
Origin: Nepal to Burma

227 An evergreen tree with wide, foot-long leaves, growing to 90 ft. in height. It is widely cultivated throughout the tropics and, as a young plant, as an indoor plant around the world. It is now being replaced, however, in popularity by the cultivar *F. e.* 'Decora' (also illustrated), with leaves that are darker green and much broader. Zone 10

224 *Ficus auriculata*

225 *Ficus benghalensis*

226 *Ficus benjamina*

227 *Ficus elastica*

227A *Ficus elastica*

28   *Ficus lyrata*

228A   *Ficus lyrata*

29   *Ficus retusa* var. *nitida*

230   *Ficus rubiginosa*

30A   *Ficus rubiginosa*

Ficus lyrata (F. pandurata)   Fiddleleaf Fig
Family: Moraceae
Origin: Tropical Africa
228   This beautiful, evergreen tree has large, lush, dark green foliage and grows to about 20 ft. with a 30 ft. spread. The leaves are about 15 in. long and 6–8 in. wide, shaped like a fiddle. It is widely grown in the tropics but as a house plant in colder areas.   Zone 10

Ficus retusa var. *nitida (F. microcarpa*   Indian Laurel Fig
var. *nitida)*
Family: Moraceae
Origin: Southern Asia
229   This is a beautiful, rounded tree with a silver-gray trunk, upright branches, and clean, dark green foliage. It is one of the most common of the *Ficus* (fig) species grown, reaching 75 ft. in height and as broad. It can stand pruning at any time and so lends itself to formal shaping and tub culture, both indoors and out.   Zone 10

Ficus rubiginosa (F. australis)   Little-leaf Fig
                                                 Port Jackson Fig
Family: Moraceae   Rusty Fig
Origin: Eastern Australia
230   An excellent, small, evergreen shade tree to 20 ft. with 4 in. long, leathery leaves, the undersides of which are a rusty brown. It needs a moist soil.   Zone 10

231   *Filicium decipiens*

Filicium decipiens   Fern Tree
                                                              Niroli

Family: Sapindaceae
Origin: India
231   A medium-sized, evergreen tree with handsome, fern-like,
dark green foliage. The leaves are composed of a dozen or so
narrow, stemless leaflets, 4–6 in. long. The 5-parted flowers are
borne at the leaf axils in panicles, 6 in. long. The plant is
monoecious, that is, bearing both male and female
flowers.   Zone 10

232   *Gliricidia sepium*

Gliricidia sepium (G. maculata)   Madre
                                               Mother of Chocolate
Family: Leguminosae   Nicaraguan Cocoa·
Origin: Tropical America   Shade
232   A deciduous tree to 30 ft., with drooping branches and 8–1(
in. long racemes of rose-pink, fragrant flowers. The leaves, seeds
and bark are poisonous if eaten.   Zone 1(

Grevillea robusta   Silk Oak
Family: Proteaceae
Origin: Australia
233   This tall, narrow, evergreen tree grows to 100 ft. or more. Its
fern-like foliage is a deep green. In the spring and summer, it bears
yellow-orange flowers. It is grown for timber in the tropics and
also is used as a street tree and in landscaping.   Zone 9

233   *Grevillea robusta*

233A   *Grevillea robusta*

234 *Harpephyllum caffrum*

235 *Hymenosporum flavum*

arpephyllum caffrum Kaffir Plum

amily: Anacardiaceae
rigin: South Africa

34 An evergreen tree to 30 ft. grown as an ornamental. It has
ce foliage, greenish-white flowers, and dark red fruit used for
aking jelly. Zone 10

ymenosporum flavum Sweet Shade

amily: Pittosporaceae
rigin: Australia

35 This delightful, evergreen, pyramidal tree, sometimes
own as a large bush, grows to a height of 60 ft. or more. For about
wo months in early summer it blooms profusely with sweet-
ented sprays of creamy yellow flowers that deepen to chrome-
ellow. This brittle tree should be planted in a quiet area and
aked. It does not like to dry out. Zone 10

caranda mimosifolia (J. ovalifolia; Jacaranda
gnonia caerulea)

amily: Bignoniaceae
rigin: Northwestern Argentina and Brazil

36 This is a prized 60 ft. ornamental in areas where it can be
own. While sometimes evergreen, it usually is deciduous for a
ry short time just as it bursts into bloom. Its spectacular display
large, conspicuous clusters of lavender-blue sprays is short-
ed. In northern Argentina it commonly is used for lumber. The
me is a Latinized form of a native Brazilian word. Zone 10

236 *Jacaranda mimosifolia*

236A *Jacaranda mimosifolia*

Kigelia pinnata Sausage Tree

Family: Bignoniaceae
Origin: Tropical Africa

237 This large tree to 50 ft. spreads to a good size and attracts
attention because of its sausage-like fruit, often up to 1 in. long and
4 in. in diameter. The fruit is not edible but is used medicinally. The
rough leaves fall once a year but are replaced in 10 days. The
flowers are wine-red. Zone 10

237 *Kigelia pinnata*

238   *Lagerstroemia indica*

Lagerstroemia indica (L. elegans)   Crape Myrtle
Family: Lythraceae
Origin: China
238   A deciduous shrub or small tree to 30 ft. This extremely
popular tree is to the people in warm areas what the lilac is to those
in colder parts. During the late summer it is covered with a profusion of red, pink, lavender, or white flowers. It requires full sun
and, in cooler areas, is subject to mildew.   Zone 7

Lagerstroemia speciosa (L. flos-reginae)   "Pride of India"
                                                    Queen's Crape
Family: Lythraceae   Myrtle
Origin: India, China, New Guinea, Australia
239   This deciduous tree to 60 ft. or more has long, lanced-shaped
leaves, which turn vivid red and fall after cold spells. The flowers,
pink or lavender but rarely white, are borne in sprays 12–15 in.
long and up to 5 in. wide—a beautiful tree when in full bloom in
June and July. It is widely planted in the tropics where the wood is
used for railroad ties and general construction. The roots, seeds,
leaves, and bark are used for medicinal purposes.   Zone 10

238A   *Lagerstroemia indica*

239   *Lagerstroemia speciosa*

239A   *Lagerstroemia speciosa*

240A   *Litchi chinensis*

tchi chinensis (Nephelium litchi)   Leechee
                                                        Litchi
mily: Sapindaceae   Lychee
rigin: China

40   This evergreen tree grows to 30 ft. and is widely planted in th the Orient and the tropics. It has a dense, rounded crown with nall, insignificant, whitish flowers, which grow in panicles about t. long in June and July. These are followed by edible nuts borne large, red clusters.   Zone 10

Macadamia integrifolia (M. ternifolia)   Queensland Nut
                                                       Smooth-Shelled
Family: Proteaceae   Macadamia
Origin: Australia

241   A rapidly growing, evergreen tree to 60 ft., it bears very hard-shelled, delicious nuts, and is grown commercially in Florida and Hawaii. In warm areas it is a good garden tree; however, if one wants to be sure of good nuts, the grafted variety of either this species or *M. tetraphylla* should be purchased.   Zone 10

241A   *Macadamia integrifolia*

1   *Macadamia integrifolia*

242   *Mangifera indica*

242A   *Mangifera indica*

Mangifera indica   Mango
Family: Anacardiaceae
Origin: India; Burma; Malaysia

242   This is one of the most important fruit trees grown in tropical countries. It is a wide-branching, erect tree to 90 ft. The leathery leaves are up to 1 ft. across, the new leaves red or yellowish. The flowers in July are small and are followed by fragrant, 2–4 in. long fruit with a peach-like flesh. The tree is widely used for heavy shade, as well as for its fruit. Some people are allergic to the sap, which may cause skin irritation.   Zone 10

Manilkara zapota (M. zapotella;   Chicle Tree
Achras zapota; Saphota achras)   Naseberry
Family: Sapotaceae   Sapodilla
Origin: Mexico; Central America

243   A densely foliaged, 100 ft., evergreen tree with glossy, 3– in., dark green leaves. The flowers are small and whitish and produce a brown fruit, about 2–4 in. long and 2 in. wide. The trees can be tapped every two years, yielding about 50 quarts of a milky latex, chicle, the original base for chewing gum.   Zone 1

243   *Manilkara zapota*

243A   *Manilkara zapota*

244A   *Melaleuca quinquenervia*

44   *Melaleuca quinquenervia*

Melaleuca quinquenervia   Cajeput Tree
(M. leucadendra)   Paperbark Tree
Family: Myrtaceae   Swamp Tea Tree
Origin: New Caledonia; New Guinea

44   An excellent tree to 25 ft. or a large shrub, widely used in
armer areas, although it has become a pest in southern Florida,
unning wild in some places. It has a whitish, shaggy bark, 8 in.
aves, and distinctive, light-colored flowers. It is a common cause
f respiratory problems at those times of the year when in
loom.   Zone 9

Melia azedarach (M. australis;   Chinaberry Tree
M. japonica; M. sempervirens)   Persian Lilac
Family: Meliaceae   Pride of India
Origin: Asia; now naturalized throughout tropical and sub-
tropical regions

245   A large, deciduous tree growing to 40 ft., it now is
naturalized in the New World. In the spring the tree is covered
with lilac-like, sweet-scented flowers. The ½ in. diameter seeds
remain on the tree for a long period. Birds are believed to eat the
fruit but it is said to be poisonous to human beings.   Zone 8

245A   *Melia azedarach*

45   *Melia azedarach*

246   *Messerschmidia argentea*

246A   *Messerschmidia argentea*

Messerschmidia argentea (Tournefortia   Tahinu
argentea)   Tree Heliotrope

Family: Boraginaceae
Origin: Tropical Indian Ocean Islands

246   This is a small, umbrella-shaped, evergreen tree to 20 ft.,
with clusters of small, white flowers. The leaves are large, 4–9 in.
long and about 4 in. wide, thick, leathery, and covered with silky,
white hairs. They are clustered at the ends of the branches. The
plant is salt-tolerant.   Zone 10

Metrosideros collina   Ohi'a-Lehu

Family: Myrtaceae
Origin: Hawaii

247   An evergreen tree to 100 ft. or shrub that grows at eleva-
tions between 1,000 and 9,000 ft. It is the first tree to be foun
growing after a volcanic eruption. The flowers are numerou
about 1 in. long with bright red stamens, similar to those of th
bottlebrush. There is a legend attached to this tree: If the blosson
are picked, it will rain.   Zone

247   *Metrosideros collina*

247A   *Metrosideros collina*

248 Monodora myristica

249 Oreocallis pinnata

Monodora myristica   African Nutmeg
Family: Annonaceae
Origin: Africa
248 An evergreen tree to 75 ft. with very interesting flowers drooping from the lower side of the branches. The flowers are fragrant, pale yellow with red spots, and up to 10 in. long and 6 in. wide. The fleshy seeds are aromatic and sometimes used as a substitute for nutmeg.   Zone 10

Oreocallis pinnata (Embothrium   Firewheel Tree
wickhamii)
Family: Proteaceae
Origin: Eastern Australia
249 A tree to 40 ft. with 8–15 in. long, pinnate leaves and bright red, pinwheel-like flowers. Very rare.   Zone 10

Pachira aquatica (P. grandiflora;   Guiana Chestnut
P. macrocarpa)   Provision Tree
Family: Bombacaceae   Wild Cocoa Tree
Origin: Tropical Mexico to northern South America
250 This spreading, evergreen tree to 60 ft. has compound leaves up to 12 in., which spread out like fingers of the hand. The fragrant flowers, which last less than a day, have 14 in. long, white petals with a red tip. The fruit is borne in a pod, 15 in. long and 5 in. across, filled with the seeds, which are pulp-covered and eaten raw.   Zone 10

250 Pachira aquatica

250A Pachira aquatica

251   *Pandanus utilis*

Pandanus utilis   Common Screw Pine

Family: Pandanaceae
Origin: Madagascar

251   This tree to 60 ft. is widely used in the tropics. The leaves, up to 6 ft. long, are used by the natives for weaving baskets and mats. The fruit is round, about 6 in. across, originally green and yellowing as it ripens, with a small amount of edible pulp.   Zone 10

252   *Parkinsonia aculeata*

Parkinsonia aculeata   Jerusalem Thorn
                                              Mexican Palo Verde

Family: Leguminosae
Origin: Tropical America

252   This spreading, dainty, deciduous tree to 20 ft., with tiny leaves, covers itself with very small, pea-shaped, yellow flowers during the winter and spring. It will grow equally well in desert conditions and along the coast as it is not only extremely drought-resistant but also resists coastal winds (as well as small boys who might consider climbing it but are deterred by its 1 in. spines). It is widely used in dry, tropical areas as well as hot areas in the interior.   Zone 9

252A   *Parkinsonia aculeata*

253   *Pimenta dioica*

Pimenta dioica (P. officinalis)   Allspice
                                                        Pimento

Family: Myrtaceae
Origin: West Indies; Central America

253   An evergreen tree to 40 ft. The leaves are picked green and dried to produce the combined flavors of nutmeg, cloves, and cinnamon. The tree is best grown in a hot, dry climate. Allspice is the dried, unripe fruit.   Zone 10

54 *Pithecellobium dulce*

255 *Pittosporum phillyraeoides*

Pithecellobium dulce (Inga dulcis)   Huamuchil
Family: Leguminosae   Opiuma
Origin: Mexico; Central America

254 This very interesting evergreen tree has both green and white leaves. It grows to 60 ft., with 2 in. leaves in pairs on very spiny twigs. The flowers are small and white and develop into 4–6 in. pods. Unfortunately, thorny seedlings scattered by the birds appear almost everywhere in tropical gardens.   Zone 10

Pittosporum phillyraeoides   Desert Willow
      Narrow-leafed
Family: Pittosporaceae   Pittosporum
Origin: Australia

255 A shrub or small tree to 30 ft. with narrow, 4 in. long leaves and pendulous branches. Fragrant, yellow flowers are produced in the spring.   Zone 10

Pittosporum rhombifolium   Queensland Pittosporum
Family: Pittosporaceae
Origin: Eastern Australia

256 Distinctive, diamond-shaped, light green foliage makes this an attractive tree to 80 ft. in the garden. Clusters of small, orange seed pods are present the year around.   Zone 10

256 *Pittosporum rhombifolium*

256A *Pittosporum rhombifolium*

257   *Pittosporum undulatum*

Pittosporum undulatum   Mock Orange
Family: Pittosporaceae   Victorian Box
Origin: Australia

257   An evergreen shrub or tree to 40 ft. The leaves are up to 6 in. long. Fragrant, ½ in., yellowish flowers grow in clusters, in turn becoming clusters of orange seeds.   Zone 9

258   *Plumeria obtusa* 'Singapore'

258A   *Plumeria obtusa* 'Singapore'

Plumeria obtusa 'Singapore'   Frangipani Tree
(*P. emarginata*)   Pagoda Tree
Family: Apocynaceae
Origin: Bahamas; Greater Antilles

258   This lovely species is the most widely planted *Plumeria* in Hawaii. The evergreen foliage is large, clean, and shiny, with clusters of creamy white flowers most of the year.   Zone 10

Plumeria rubra   Frangipani
Family: Apocynaceae   Temple Tree
Origin: Tropical America

259   Another small tree, to 15 ft., with pink to red flowers. The plant sheds its leaves for a short time in winter and comes into bloom in early spring. Shortly afterward, the 3×6 in. leaves appear. The flowers continue until the leaves drop in the fall.   Zone 10

259   *Plumeria rubra*

259A   *Plumeria rubra*

259B   *Plumeria rosea*

260   *Podocarpus gracilior*

Podocarpus gracilior   Fern Pine

Family: Podocarpaceae
Origin: Tropical Africa

260   This is one of the numerous species of *Podocarpus* native to Africa, where it is used as a timber tree for fine furniture. It is at home in tropical areas growing to 75 ft. or more as well as in temperate zone gardens to 25°F. Although it does not look like it, it is a conifer. It is widely used in Zones 9 and 10 as a landscape plant and as a container plant, both indoors and outside.   Zone 10

Podocarpus macrophyllus   Japanese Yew Pine
(*P. longifolius*)

Family: Podocarpaceae
Origin: Japan

261   This 45 ft., evergreen tree is more tolerant of heat and drought than other species of *Podocarpus*. It will grow indoors in containers or outdoors, where it often is pruned to topiary, hedge, or tree form. It usually is grown, however, as a narrow, trimmed specimen on either side of a doorway.   Zone 8

Pseudobombax ellipticum (*Bombax*   Shaving Brush Tree
ellipticum; Pachira fastuosa)

Family: Bombacaceae
Origin: Tropical America; West Indies

262   A spreading, deciduous tree to 30 ft. which blooms in the spring before the leaves appear. The flowers are 3–6 in. long, usually pink but sometimes white, and remind one of a shaving brush.   Zone 10

261   *Podocarpus macrophyllus*

262   *Pseudobombax ellipticum*

262A   *Pseudobombax ellipticum*

263   *Ravenala madagascariensis*

263A   *Ravenala madagascariensis*

264   *Samanea saman*

264A   *Samanea saman*

265   *Sambucus mexicana* var. *bipinnata*

Ravenala madagascariensis   Travelers Palm
Family: Strelitziaceae   Travelers Tree
Origin: Madagascar

263   This is a highly valued ornamental tree that grows to 30 ft.
and suckers from the base so a clump eventually will form. It has a
palm-like trunk and banana-shaped leaves, giving the appearance
of a fan. Water collects in the hollow base of the leaves providing a
source of liquid for travelers, which accounts for its common
name.   Zone 10

Samanea saman (Pithecellobium saman)   Monkey Pod Tree
Family: Leguminosae   Rain Tree
Origin: Central America; West Indies

264   This beautifully shaped tree grows to 80 ft. with a spread up
to 100 ft. It is fast-growing and so is not suitable for a small garden
but is ideal for parks and open spaces. The fern-like leaves close on
cloudy days and in late afternoon and drop in February and March.
The wood is used for the lovely bowls one sees in gift
shops.   Zone 10

Sambucus mexicana var. *bipinnata*   Elderberry
Family: Caprifoliaceae
Origin: Mexico

265   A tree to 50 ft. bearing a large number of flat-topped flowerheads in clusters, composed of many small, white flowers. These
develop into clusters of small, purple-black berries, which are
edible and often made into jelly or wine. They are also a great
favorite of birds.   Zone 8

266 *Saraca indica*

266A *Saraca indica*

Saraca indica   Asoka Tree
Family: Leguminosae   Sorrowless Tree
Origin: India to the Malay Peninsula
266   An evergreen tree to 30 ft. The orange-red flowers deepen with age and are fragrant. It is a favorite ornamental in India, where the tree is considered sacred by the Hindus who use the flowers as temple offerings.   Zone 10

267 *Schinus terebinthifolius*

267A *Schinus terebinthifolius*

Schinus terebinthifolius   Brazilian Pepper Tree
Family: Anacardiaceae   Christmas Berry Tree
Origin: Brazil
267   An evergreen tree or shrub to 20 ft., with dark green leaves and greenish flower clusters in summer, followed by dense clusters of bright red berries at Christmastime. It is used as a decorative landscape tree in Zones 9 and 10. In Florida it is called the "Christmas Berry" and has escaped, forming jungles from the seeds dropped by birds and is crowding out the native vegetation. It is a common source of skin irritation as well as a cause of respiratory problems.   Zone 10

268   *Spathodea campunulata*

268A   *Spathodea campunulata*

269A   *Spathodea campunulata aurea*

269   *Spathodea campunulata aurea*

270   *Stenocarpus sinuatus*

Spathodea campunulata   African Tulip Tree
                                                                   Fire Tree
Family: Bignoniaceae   Flame of the Forest
Origin: Tropical Africa   Fountain Tree

268   This tree is seen throughout the tropical world. It is an out-standing and colorful evergreen. It fiery red flowers can be recognized from a great distance, as it grows to 50 ft. or more and as wide. The buds hold water so youngsters use them for water pistols. It is truly one of the most outstanding trees in the world.   Zone 10

Spathodea campunulata aurea   African Tulip Tree

Family: Bignoniaceae
Origin: Tropical Africa

269   This bright yellow to orange-yellow sport also is found but the trees usually are smaller in size than *S. campunulata.*   Zone 10

Stenocarpus sinuatus   Firewheel Tree
Family: Proteaceae   Wheel of Fire
Origin: New South Wales and Queensland

270   This evergreen is indeed a spectacular tree. The large, deeply cut, dark green leaves are very attractive, and in late summer, when the rich scarlet flowers in the form of a pinwheel appear, the combination is a sight to remember. The tree grows to a height of 100 ft.   Zone 10

71   Syzygium malaccense

272   Tabebuia caraiba

273   Tabebuia donnell-smithii

274   Tabebuia rosea

274A   Tabebuia rosea

yzygium malaccense (Eugenia
alaccensis; Jambosa malaccensis)   Malay Apple
                                                 Pomeral Jambos
amily: Myrtaceae   Rose Apple
rigin: Malay Peninsula

71   This is one of the most beautiful tropical trees, growing to 40
., and is suitable for growing only in hot, frost-free areas. The
aves are 6–12 in. long, dark green, and leathery. The flowers are a
nowy purplish-red. The yellowish, 2 in. long fruit is eaten raw or
reserved.   Zone 10

abebuia caraiba (T. argentea;   Golden Bell
ecoma argentia)   Silver Tree
amily: Bignoniaceae   Tree of Gold
rigin: Argentina; Paraguay

72   A tree to 20 ft., it is spectacular when in bloom in early
pring. The 5 in. long leaves are covered with dense, silvery scales.
he tubular, yellow flowers are 2–3 in. long and grow in
lusters.   Zone 10

abebuia donnell-smithii (Cybistax   Gold Tree
onnell-smithii)   Primavera
amily: Bignoniaceae
rigin: Mexico; Central America

73   One of the most beautiful deciduous trees in the world. It
rows to 75 ft., with a spreading crown of masses of bright yellow,
ell-shaped flowers, which last for about 2 months, beginning in
March.   Zone 10

abebuia rosea (T. pallida;   Pink Poui-Rosea
 pentaphylla)   Pink Trumpet Tree
amily: Bignoniaceae
rigin: Mexico to Venezuela

74   This shrub or small tree, from 20–50 ft., is grown for its hand-
ome foliage and profusion of attractive, pink flowers, which
ppear frequently throughout the year.   Zone 10

275   *Tabebuia serratifolia*

Tabebuia serratifolia   Trumpet Tree
Family: Bignoniaceae   Yellow Poui
Origin: West Indies to Bolivia

275   A briefly deciduous tree to 50 ft., it is one of the showiest flowering trees in the tropics. The large clusters of tubular, yellow flowers usually arrive before the leaves and last after they are gone.   Zone 10

276   *Taxodium distichum*

Taxodium distichum   Bald Cypres
Family: Taxodiaceae   Swamp Cypres
Origin: Southeastern United States

276   A deciduous conifer tree of swamps and streamside, it ofte grows to more than 100 ft. It is striking in the fall when the foliag turns bronze, but it also is beautiful in the spring when the ne foliage appears. When growing in swamps, it sends up "knees" t get air.   Zone 5

277   *Taxodium mucronatum*

Taxodium mucronatum (T. mexicanum)   Montezuma Cypress
Family: Taxodiaceae
Origin: Mexico

277   This little-known, coniferous tree is similar to *T. distichum*, but the leaves are shorter and the cones larger. It is deciduous in colder climates.   Zone 8

277A   *Taxodium mucronatum*

Tectona grandis   Teak
Family: Verbenaceae
Origin: India; Burma
278   This deciduous trees grows to 150 ft. Tiny, white flowers are borne in clusters. There are many managed forests in Sri Lanka and Thailand where the trees are grown for the wood, which is used in fine furniture, ships, etc.   Zone 10

Terminalia catappa   False Kamani
Tropical Almond
Family: Combretaceae
Origin: East Indies
279   An 80 ft., deciduous tree grown in many tropical countries for shade, timber, and the edible nuts. It grows well near sandy beaches. The leaves, 6–12 in. long, are bright green but there usually is a scattering of bright red leaves before they fall, which remain for some time but are replaced with new leaves soon afterward.   Zone 10

Tristania conferta   Brisbane Box
Family: Myrtaceae
Origin: Australia
280   An attractive, roundheaded, evergreen tree to 50 ft. in cultivation but much larger in the wild. The dark green foliage contrasts nicely with the reddish-brown bark. It likes plenty of heat and withstands drought.   Zone 9

278   *Tectona grandis*

278A   *Tectona grandis*

279   *Terminalia catappa*

280   *Tristania conferta*

281   *Tupidanthus calyptratus*

Tupidanthus calyptratus   Mallet Flower
Family: Araliaceae
Origin: India to Cambodia
281   A large, evergreen shrub or small tree to 15 ft. with large, fan-like, glossy green leaves. It resembles the *Schefflera* but develops a multiple trunk, which makes for a denser and broader plant. It is an excellent indoor plant.   Zone 10

VINES

282   *Allamanda cathartica*

282A   *Allamanda cathartica*

Allamanda cathartica   Golden Trumpet
Family: Apocynaceae   Yellow Allamanda
Origin: Brazil

282   A vigorous, evergreen vine to about 50 ft., often used as a
heavy bank cover or can be cut back to form a mounded shrub. The
flowers are bright yellow with a paler throat, usually 3–4 in. across
and 1½ in. long. The leaves are about 5 in. long and shiny. Both the
leaves and bark are cathartic, as well as irritating to the skin of some
people. The plant enjoys poor soil.   Zone 10

283   *Allamanda neriifolia*

284   *Allamanda violacea*

Allamanda neriifolia   Bush Allamanda
Family: Apocynaceae
Origin: Brazil

283   An evergreen vine, often grown as a bush, to 4–5 ft., which
flowers the year round. The yellow flowers, with an orange-
striped throat, are smaller than those of the more common species
A. cathartica. The plant enjoys poor soil and neglect. The leaves are
poisonous when eaten.   Zone 10

Allamanda violacea   Purple Allamand
Family: Apocynaceae
Origin: Brazil

284   A semi-climbing, shrub-type vine with reddish-purp
flowers, 2–3 in. across and very showy. Will grow in full sun an
does best when grafted onto *A. cathartica.*   Zone 1

Antigonon leptopus   Chain of Love Vine
                                                            Coral Vine
                                                      Mexican Creeper

Family: Polygonaceae
Origin: Mexico
285   This vine will climb a fence or tree to 40 ft. and is covered with small, bright pink, heart-shaped flowers. It begins to bloom in early spring and lasts throughout the summer.   Zone 10

285   Antigonon leptopus

Aristolochia elegans   Calico Flower

Family: Aristolochiaceae
Origin: Brazil
286   A slender, woody vine with heart-shaped leaves to 4 in. long and flowers with a calyx tube inflated to 1½ in. and about 3 in. across. The flowers are rich purplish-brown, marked with white inside.   Zone 10

Aristolochia gigantea (A. grandiflora)   Pelican Flower

Family: Aristolochiaceae
Origin: Brazil
287   A climbing, herbaceous vine to 10 ft., with heart-shaped, pointed leaves up to 10 in. long. The duck-shaped, inflated flower bud develops to a flower from 6–20 in. long with a 3 in. tail. The flower is mottled purple on the outside with a white and purple throat. Flowers in early summer.   Zone 10

286   Aristolochia elegans

287   Aristolochia gigantea

288   Bauhinia corymbosa

289   Beaumontia grandiflora

289A   Beaumontia grandiflora

Bauhinia corymbosa   Phanera

Family: Leguminosae
Origin: Southeast Asia

288   This vine grows to 15 ft. and is ideal for an arbor. It has small leaves on tendril-bearing branches. Free-flowering in clusters of light pink flowers that are veined white or pink in the summer.   Zone 10

Beaumontia grandiflora   Easter-Lily Vine
                                                            Herald's Trumpet

Family: Apocynaceae
Origin: Himalayas

289   This spectacular, evergreen vine blooms in the spring, with large, bell-shaped, fragrant, white flowers, 4 in. in length, and heavy, shiny leaves, up to 8 in. long. The plant is very woody so it requires a stout support to carry the weight.   Zone 10

BOUGAINVILLEA

290 Bougainvillea brasiliensis

Bougainvillea
Family: Nyctaginaceae
Origin: Brazil
290–290F These vigorous and showy plants come in many colors, with both single and double forms. The leaf bracts give the color and hide the tiny flowers. They demand a well-drained soil and do not like much water. Reflected heat in a hot, desert-like area is ideal.   Zone 10

290A Bougainvillea spectabilis 'San Diego Red'

290B Bougainvillea spectabilis 'Orange King'

290C Bougainvillea spectabilis 'La Jolla'

290D Bougainvillea spectabilis 'Hong Kong Red'

290E Bougainvillea (Mixed Varieties)

290F Bougainvillea spectabilis 'Barbara Karst'

291   *Campsis* × *tagliabuana* 'Madame Galen'

Campsis × *tagliabuana* 'Madame   Trumpet Creeper
Galen'

Family: Bignoniaceae
Origin: Garden Hybrid

291   A vigorous, deciduous, climbing vine that will cling to most surfaces and produce loose, arching sprays of trumpet-shaped, red flowers. Excellent screen. It also is grown as a shrub in clump form.   Zone 6

292   *Clerodendrum thomsonae*

Clerodendrum thomsonae   Bleeding Heart Vine
Family: Verbenaceae   Glory Bower
Origin: Tropical West Africa

292   A twining, evergreen shrub-vine to 12 ft. with narrow, 5 in. long leaves and very unusual clusters of flowers. They are white with a brilliant red calyx and about 1 in. across. They often are grown as a pot plant in the colder parts of the world.

A legend connected with this plant says that it arose from the tears of a maiden who had been deserted and then ran away. Where her tears fell, a plant bearing blossoms in the shape of bleeding hearts sprang up.   Zone 10

293   *Clystosoma callistegioides*

Clytostoma callistegioides (*Bignonia*   Lavender Trumpet
callistegioides; B. speciosa; B. violacea)   Vine
Family: Bignoniaceae   Love-charm
Origin: Southern Brazil; Argentina

293   A strong-growing, evergreen vine, which will grow in either sun or partial shade. Large sprays of large clusters of 3–4 in., trumpet-shaped, lavender flowers cover the foliage from April to August.   Zone 10

293A   *Clystosoma callistegioides*

294 *Congea tomentosa*

294A *Congea tomentosa*

Congea tomentosa   Pink Shower Orchid
Family: Verbenaceae   Shower-of-Orchids
Origin: Burma; Thailand
294 A climbing vine, often grown as a shrub, to 10 ft. with long, loose sprays of spreading branches. Clusters of 3 in. long flower bracts that open white and then turn to a pinkish-lavender cover the branches.   Zone 10

Cryptostegia grandiflora   Rubber Vine
Family: Asclepiadacea
Origin: Africa
295 A strong-growing, vining shrub to about 6 ft. The lilac, bell-shaped flowers, 2 in. across, eventually turn pink. It blooms over a long period in the summer and fall.   Zone 10

295 *Cryptostegia grandiflora*

296   *Distictis buccinatoria*

296A   *Distictis buccinatoria*

Distictis buccinatoria (Bignonia cherere;   Blood Trumpet Vine
Phaedranthus buccinatorius)

Family: Bignoniaceae
Origin: Mexico

296   This popular, evergreen vine has dark green, oblong leaves up to 4 in. long. Clusters of 4 in., trumpet-shaped, bright red flowers stand out and cover the vine.   Zone 10

Epipremnum aureum (Pothos aureus;   Golden
Raphidophora aurea; Scindapsus aureus)   Philodendron
                                              Golden Pothos
Family: Araceae
Origin: Southeast Asia

297   In colder areas, this plant usually is sold under the name Golden Pothos for indoor use. The leaves are no more than 3 in. wide. In their native habitat, the evergreen plants with little, new leaves remain small while on the ground but, when they find a tree or structure to climb on, the leaves grow in size up to 2 ft. across.   Zone 10

Hoya carnosa   Honey Plant
                                                 Wax Plant
Family: Asclepiadaceae
Origin: Southern China to Australia

298   This climbing, evergreen vine produces fragrant, waxy, star-shaped, pink-and-white blooms in clusters. In frost-free, warm regions, it can be grown in partial shade with very little water. Hummingbirds love the rich nectar of the flowers. Spring- and summer-flowering.   Zone 10

297   *Epipremnum aureum*

298   *Hoya carnosa*

298A   *Hoya carnosa*

99 *Ipomoea horsfalliae*

Ipomoea horsfalliae (I. briggsii)   Prince Kuhio Vine
Family: Convolvulaceae
Origin: West Indies
99   A large, glabrous, woody, evergreen vine, it is commonly cultivated in the tropics and in greenhouses worldwide. It is very showy, with masses of magenta-crimson (sometimes pale purple) flowers, 2–3 in. in size.   Zone 10

299A   *Ipomoea horsfalliae*

00   *Ipomoea pes-caprae*

Ipomoea pes-caprae (I. bibol)   Beach Morning Glory
        Railroad Vine
Family: Convolvulaceae
Origin: Tropical America; Africa
300   A creeping vine with rope-like stems up to 50 ft., found on sandy beaches. The 4 in. flowers are purple to dull pink. This very good sand plant is salt-tolerant and grows on the beaches even below the high-tide line.   Zone 10

Ipomoea purpurea (I. mexicana;   Common Morning
Pharbitis purpurea)   Glory
Family: Convolvulaceae
Origin: Tropical America
301   This annual plant is widely naturalized throughout the tropics and is a very common garden vine. The colors of the cultivars are purple, pink, blue, or white.   Zone 10

301   *Ipomoea purpurea*

302   *Ipomoea tuberosa*

302A   *Ipomoea tuberosa*

302B   *Ipomoea tuberosa*

303   *Jasminum mesnyi*

304   *Jasminum multiflorum*

304A   *Jasminum multiflorum*

Ipomoea tuberosa (Merremia tuberosa; Operculina tuberosa)   Wood Rose
                                                           Yellow Morning
                                                                    Glory
Family: Convolvulaceae
Origin: Tropical America

302   This vine is found almost everywhere in the tropics. The 2 in. flowers are yellow-orange. When dried, they are brownish and almost transparent. The dried seed pods frequently are used in flower arrangements. The plant develops a tuber from which it is renewed from year to year.   Zone 10

Jasminum mesnyi (J. primulinum)   Primrose Jasmine
Family: Oleaceae
Origin: Western China

303   A fast-growing, evergreen, vining-shrub to 10 ft., with small, bright green leaves and covered with double, lemon-yellow flowers. It needs the support of a fence or trellis to reach its full height.   Zone 8

Jasminum multiflorum (J. pubescens)   Star Jasmine
Family: Oleaceae
Origin: India; China

304   This evergreen jasmine is grown as both a vine and a mounded shrub. The slightly fragrant, white flowers are borne almost all year. It can stand heavy pruning and will still bloom.   Zone 6

05   *Jasminum nitidum*

306   *Jasminum polyanthum*

asminum nitidum (J. magnificum)   Angel Wing Jasmine
                                             Windmill Jasmine

amily: Oleaceae
Origin: Admiralty Islands
05   A semi-vining, spreading, evergreen shrub with large,
ragrant, windmill-like, white flowers that bloom during the sum-
ner.   Zone 10

asminum polyanthum   Pink Jasmine

amily: Oleaceae
Origin: Western China
06   A vigorous, climbing, sometimes deciduous, evergreen vine
vith delicate, lacy foliage. In late spring, it produces panicles of
oink buds, which open to white flowers. Beautifully fragrant.
                                                     Zone 8

307   *Lonicera hildebrandiana*

307A   *Lonicera hildebrandiana*

Lonicera hildebrandiana   Giant Burmese
                                               Honeysuckle

Family: Caprifoliaceae
Origin: Burma; China
307   A fast-growing, evergreen vine, it is one of the largest in the
honeysuckle group. It has large, shiny leaves and clusters of giant,
6–7 in. long fragrant, cream to yellow flowers, which bloom over a
long period.   Zone 10

308   *Mandevilla splendens*

Mandevilla splendens (Dipladenia splendens)   Pink Mandevilla

Family: Apocynaceae
Origin: Brazil

308   An evergreen, glabrous, woody, twining vine, the plant must be supported on a fence or trellis. It needs a rich, well-drained soil with plenty of water during the summer. Even in Zone 10 it should be grown in a warm, sheltered position. Large, pink flowers cover the plant in mid- to late summer.   Zone 10

308A   *Mandevilla splendens*

309   *Monstera deliciosa*

Monstera deliciosa (Philodendron pertusum)   Split-leaf Philodendron
Swiss Cheese Plant

Family: Araceae
Origin: Mexico; Central America

309   This vine grows to 30 ft. in its native habitat, with leaves up to 3 ft. wide. It usually is used in tropical landscapes, however, as a foundation plant kept at 6 ft. In colder areas, it can be grown successfully indoors in a well-lighted room, facing the light; otherwise the leaves will turn to face the light.   Zone 10

309A   *Monstera deliciosa*

310 *Mucuna bennettii*

Mucuna bennettii   New Guinea Creeper
Family: Leguminosae   Red Jade
Origin: New Guinea

310   A woody climber, it is very rare and grown only occasion-ally in tropical gardens. The 3 in. long, orange or red flowers grow in very showy, pendulant clusters.   Zone 10

310A *Mucuna bennettii*

Pandorea jasminoides (Bignonia   Bower Plant
jasminoides; Tecoma jasminoides)

Family: Bignoniaceae
Origin: Australia

311   A fast-growing vine to 20 ft. or more, it does not like wind, drying out, or frost. The foliage is glossy and dark green. The 2 in., tubular flowers are rose-pink with a white throat and bloom from June to October.   Zone 10

Passiflora × alatocaerulea (P. alata ×   Passion Flower
P. caerulea; P. pfordtii)

Family: Passifloraceae
Origin: Garden origin

312   The large, exotic, fragrant flowers on this fast-growing, hybrid vine measure 3–4 in. across. They are used in the manufac-ture of perfumes, as well as in landscaping to cover fences and trellises. The plant does not produce fruit.   Zone 10

Passiflora coccinea   Red Granadilla
Family: Passifloraceae   Red Passion Flower
Origin: Venezuela to Bolivia

313   This showy, winter-flowering vine produces edible, orange or yellow fruit but requires pollination. The 3–5 in. flowers are deep red to orange.   Zone 10

311 *Pandorea jasminoides*

312 *Passiflora × alatocaerulea*

313 *Passiflora coccinea*

314   *Passiflora edulis*

315   *Passiflora jamesonii*

315A   *Passiflora jamesonii*

316   *Petrea volubilis*

316A   *Petrea volubilis*

Passiflora edulis   **Passion Fruit**
Family: Passifloraceae   **Purple Granadilla**
Origin: Brazil

314   A vigorous vine with showy flowers, which produce edible, deep purple, 2 in. fruit used in making beverages, sherbet, and jellies. Grows in full sun.   Zone 10

Passiflora jamesonii (Tacsonia   **Pink Passion Flower**
jamesonii)

Family: Passifloraceae
Origin: Ecuador

315   This beautiful, evergreen vine flowers all summer with 3 in. blooms in lovely shades of rose to coral-red. It is an excellent, fast-growing plant with which to cover a fence or bank.   Zone 10

Petrea volubilis   **Blue Bird Vine**
Family: Verbenaceae   **Queen's Wreath**
Origin: Central America; Mexico; West Indies

316   A high-climbing, evergreen vine to 35 ft., it grows best in full sun and reflected heat. The lavender-blue flowers are found in clusters 3–12 in. long.   Zone 10

317 *Pseudocalymma alliaceum*

318 *Pyrostegia venusta*

318A *Pyrostegia venusta*

Pseudocalymma alliaceum   Garlic Vine

Family: Bignoniaceae
Origin: Guianas and Brazil

317   This vine blooms off and on during the warm months. The leaves, when crushed, emit a strong garlic odor. The flowers are large, in a lovely, bell-shaped cluster about 3 in. long, in colors of pink, lavender, or white with purplish veins.   Zone 10

Pyrostegia venusta (P. ignea; Bignonia   Flame Vine
venusta)   Sweetheart Vine

Family: Bignoniaceae
Origin: Brazil; Paraguay

318   This evergreen vine is happiest in the hottest of gardens and with reflected heat, so is especially comfortable on tile roofs. The 3 in. long, flame-orange, tubular flowers appear in December and throughout the spring. It frequently is seen in tropical and sub-tropical areas in early spring.   Zone 10

Securidaca diversifolia (Polygala   Polygala
diversifolia)

Family: Polygalaceae
Origin: West Indies to Ecuador

319   This trailing or climbing shrub to 20 ft. has thick, shining, 5 in. leaves and 2–4 in. flowering racemes of pink to purplish flowers with a keel having a yellow tip.   Zone 10

319 *Securidaca diversifolia*

320   *Senecio confusus*

321   *Solandra maxima*

321A   *Solandra maxima*

322   *Solanum jasminoides*

Senecio confusus   Mexican Flame Vine
Family: Compositae   Orange Glow Vine
Origin: Mexico

320   A beautiful, evergreen vine with bright orange flowers and thickish, glabrous leaves about 2 in. long. It is best to be careful when pruning as a skin rash may develop.   Zone 10

Solandra maxima (S. guttata;   Cup of Gold
S. hartwegii; S. nitida)   Trumpet Plant
Family: Solanaceae
Origin: Mexico

321   An evergreen, climbing shrub or vine to 20 ft. It is very fast-growing when in the right spot and sometimes will grow as much as 1 ft. a day. The flowers, which are more than 9 in. long, are cream colored when first open and then turn to deep yellow. It blooms best during the winter but does have a few flowers all year.
                                                                Zone 10

Solanum jasminoides   White Potato Vine
Family: Solanaceae
Origin: Brazil

322   A vigorous, evergreen, 16 ft. vine to about 28°F, then deciduous. It grows in full sun with small, bright foliage and masses of bluish-white, star-shaped flowers, tinged with mauve.
                                                                Zone 9

Solanum rantonnetii (Lycianthes   Blue Potato Vine
cantonei)   Paraguay Nightshade
Family: Solanaceae
Origin: Argentina to Paraguay

323   A vigorous, medium-sized, semi-vining, evergreen shrub. It bears masses of royal purple flowers from spring until fall. Often trained to a standard patio tree by nurserymen and kept to a roundheaded ball by pruning.   Zone 10

323   *Solanum rantonnetii*

324 Solanum seaforthianum

325 Solanum wendlandii

| *olanum seaforthianum* | Brazilian Nightshade |
| | Star Potato Vine |

amily: Solanaceae
Origin: South America
324 A graceful, 20 ft., evergreen vine that climbs or trails for a mited space. Its 8 in. leaves are star-like and the blue or lavender owers, up to 1 in. across, hang in clusters. The seeds are bright ed, very attractive to birds but toxic to humans.   Zone 10

Solanum wendlandii	Costa Rican Nightshade
	Divorce Vine
Family: Solanaceae	Grant Potato Vine
Origin: Costa Rica	Marriage Vine

325 This vine, to 18 ft., has stout stems and a few thorns. It is a rampant grower, with showy clusters of lilac flowers in late summer and autumn. It often is used as a ground cover.   Zone 10

326 Stephanotis floribunda

327 Stigmaphyllon ciliatum

tephanotis floribunda	Bridal Bouquet
	Floradora
amily: Asclepiadaceae	Madagascar Jasmine
Origin: Madagascar	

26 This is a popular evergreen vine, to 15 ft., which grows well ut-of-doors in hot, humid places and in greenhouses in cold reas. It has large, 4 in., leathery leaves and clusters of 2 in. long, ragrant, white flowers, which often are used in bridal ouquets.   Zone 10

Stigmaphyllon ciliatum	Brazilian Golden
	Vine
	Butterfly Vine
Family: Malpighiaceae	
Origin: Tropical America	

327 A slender, twining, woody vine, with smooth, leathery, 3 in., heart-shaped leaves and clusters of 3–6 bright golden yellow flowers. The plant must be protected from strong sunlight.   Zone 10

328   *Stigmaphyllon emarginatum*

329   *Strongylodon macrobotrys*

330   *Syngonium podophyllum*

Stigmaphyllon emarginatum   Gold Creeper

Family: Malpighiaceae
Origin: Brazil to Argentina

328   This vine is tender to frost but grows very rapidly in well-drained, sunny soil; needs protection from hot, dry winds. The foliage is bright green, the leaves 3–4 in. long and 1 in. wide. Clusters of 1 in., bright yellow flowers are produced.   Zone 10

Strongylodon macrobotrys   Green Jade Vine

Family: Leguminosae
Origin: Philippines

329   A vigorous, spectacular climber, excellent for pergolas. The unusual, jade-green flowers hang in 2–3 ft. long panicles, providing its own shade, which is needed for the roots.   Zone 10

Syngonium podophyllum (Nephthytis   African Evergree
afzelii)   Arrowhead Vin
                                              Nephthyti
Family: Araceae
Origin: Mexico to Panama

330   This perennial evergreen usually is offered to the public as potted vine for indoor use but is found in tropical areas often as ground cover, where the leaves will grow to 12 in. long and half a wide when it attaches itself to a tree. The juvenile leaves are arrow shaped and variegated with gold or white. The plant is commonly but incorrectly, offered as *Nephthytis afzelii*.   Zone 1

Tecomaria capensis (Bignonia capensis;   Cape Honeysuckl
Tecoma capensis)

Family: Bignoniaceae
Origin: South Africa

331   An evergreen vine or stiff shrub to 15 ft. with dark gree foliage and clusters of orange-red, trumpet-shaped flowers. It like full sun, even in the desert, but it also is excellent at the seashore. may be kept trimmed as a hedge.   Zone 1

331   *Tecomaria capensis*

2 *Thunbergia alata*

333 *Thunbergia fragrans*

unbergia alata   *Black-eyed Susan Vine*

mily: Acanthaceae
igin: Tropical Africa

2 A twining herb, now widely naturalized throughout the
pics, it is grown as an annual in colder areas. There is a large
mber of cultivars in colors ranging from pure white to
llow.   Zone 10

Thunbergia fragrans   Sweet Clockvine

Family: Acanthaceae
Origin: India

333 This evergreen, woody, twining vine displays fragrant,
white flowers that are up to 2 in. wide. Grown from seed, it is fast-
growing and attractive but it must be kept under control as it tends
to run wild.   Zone 10

34 *Thunbergia grandiflora*

335 *Thunbergia mysorensis*

hunbergia grandiflora   Sky Flower

amily: Acanthaceae
rigin: India

34 A vigorous, evergreen vine with large, 8 in., heart-shaped
aves and clusters of flaring, 5 in., trumpet-shaped flowers of a
elicate blue. It is too vigorous for a small garden. The major
roblem is the constant chore of removing dead blos-
oms.   Zone 10

Thunbergia mysorensis   Mysore Trumpet
                                                             Vine
Family: Acanthaceae
Origin: India

335 A fast-growing, evergreen vine with deep green foliage. The
golden flowers with a mahogany-red hood grow to about 3 ft.,
unfolding continuously. This species should be grown on a per-
gola so the flowers can hang free of the leaves. Needs summer
moisture.   Zone 10

336   *Tristellateia australasiae*

336A   Tristellateia australasiae

Tristellateia australasiae   Bagnit Vine
Family: Malpighiaceae
Origin: Southeast Asia to New Caledonia
336   The leaves of this woody vine are light yellow-green and
about 6 in. long. The 1 in., yellow flowers appear in long clusters,
with each flower having a group of short, red stamens in the
center.   Zone 10

Wagatea spicata   Candy Corn Vine
Family: Leguminosae
Origin: India
337   This large, evergreen vine has large, graceful trusses of long
flower spikes at the tip ends of long branches in late summer. Each
spike is about 8 in. long with numerous orange buds, which open
up into pen-shaped flowers.   Zone 10

337   *Wagatea spicata*

PALMS AND CYCADS

338   *Archontophoenix alexandrae*

339   *Archontophoenix cunninghamiana*

Archontophoenix alexandrae   Alexandra Palm
(Ptychosperma alexandrae)

Family: Palmae
Origin: Australia

338   A very straight, slender palm to 60 ft. The arching, 10 ft. leaves do not droop below the horizontal.   Zone 10

Archontophoenix cunninghamiana King Palm
(Ptychosperma cunninghamianum; Piccabeen Palm
Seaforthia elegans)

Family: Palmae
Origin: Australia

339   The tall, slender trunk has long, arching, light green branches. The lilac or purplish flowers grow out of the main stem, 4–5 ft. below the crown.   Zone 10

339A   *Archontophoenix cunninghamiana*

Arecastrum romanzoffianum (Cocos Queen Palm
plumosa; C. romanzoffianum)

Family: Palmae
Origin: Southern Brazil to Argentina

340   This rapidly growing palm rises to 30 ft. or more, with long, graceful, 15 ft. fronds and 36 in. flower spikes. It does well in a landscape with other plants; can be used singly or in groups.
   Zone 9

340   *Arecastrum romanzoffianum*

341   Bismarckia nobilis

342   Brahea armata

Bismarckia nobilis (Medemia nobilis)

Family: Palmae
Origin: Madagascar
341   A beautiful palm to 25 ft. or more, with gray-blue, 4 ft. fronds and 4 ft. flower spikes. It is widely planted as an ornamental in the tropics.   Zone 10

Brahea armata (Erythea armata; E. roezlii)   Gray Goddess   Mexican Blue Palm

Family: Palmae
Origin: Northwestern Mexico
342   A palm with a solitary trunk to 45 ft. but very slow growing. It is attractive in the garden because of its fine, spreading crown of silver-blue, fan-like, 15 ft. fronds and 10 ft. flower stems.
                                                                Zone 9

343   Brahea edulis

344   Butia capitata

Brahea edulis (Erythea edulis)   Guadalupe Fan Palm

Family: Palmae
Origin: Guadalupe Island
343   A robust palm to about 30 ft. with a dense, 10–12 ft. spread. The spent fronds fall, naturally cleaning the trunk.   Zone 9

Butia capitata (Cocos australis)   Hardy Blue Cocos   Jelly Palm

Family: Palmae
Origin: Brazil; Uruguay
344   This species stands more heat, frost, and drought conditions than any of the other feather palms. It has graceful, recurved, 5 ft. long, silver-blue leaf fronds. It is an excellent tub plant for a hot, sunny exposure.   Zone 9

Caryota cumingii   Fishtail Palm

Family: Palmae
Origin: Philippines

345 This "fishtail" palm grows to about 25 ft. with an 8 in. trunk, but seems to have a much denser growth habit than other *Caryota* palms.   Zone 10

Caryota mitis (C. furfuraceae;   Burmese Fan Palm
C. griffithii; C. sobolifera)   Clustered Fishtail
                               Palm

Family: Palmae
Origin: Burma; Malay Peninsula; Philippines

346 Growing from 12–40 ft. high, this tree is widely planted in the tropics as an ornamental and often is seen in greenhouse collections. It has a very thin trunk. As it starts to flower, the tree begins to die, but will continue to produce seed for a few years as it is dying. The outer coat is poisonous, causing irritation to the skin.   Zone 10

345 *Caryota cumingii*

346 *Caryota mitis*

347 *Caryota no*

348 *Chamaerops humilis*

Caryota no (C. alberti; C. rumphiana)   Fishtail Palm

Family: Palmae
Origin: Australia to the Philippines

347 This "fishtail" grows to 75 ft. with a single trunk up to 18 in. in diameter. It often is planted in the warmer areas of Zone 10 as an ornamental and is widely grown elsewhere in greenhouses as a decorative house plant.   Zone 10

Chamaerops humilis   Hair Palm
                   Mediterranean Fan Palm

Family: Palmae
Origin: Mediterranean

348 A multi-trunk fan palm from 5–10 ft. which stands out from the base, forming a thick, spreading clump. It is very slow growing and, as it gets older, the outer trunks reach for the light, then twist and grow straight, giving a handsome curve to the trunks.   Zone 9

349   *Chrysalidocarpus lutescens*

350   *Coccothrinax alta*

351   *Cocos nucifera*

351A   *Cocos nucifera*

Chrysalidocarpus lutescens (Areca   Areca Palm
lutescens)   Butterfly Palm

Family: Palmae
Origin: Madagascar

349   The plant is almost always started in a clump, and, in the tropics, will grow to 30 ft. It is widely grown in tropical and subtropical areas, both as an ornamental palm and a tub plant.   Zone 10

Coccothrinax alta   Silver Palm

Family: Palmae
Origin: Puerto Rico

350   A slender palm with a smooth, slightly ringed, gray-brown trunk. It grows to about 30 ft. Deeply divided fronds.
      Zone 10

Cocos nucifera   Coconut Palm

Family: Palmae
Origin: Tropical Pacific Islands

351   The coconut palm is one of the most beautiful palms in the world and one of the best known. It will grow to 100 ft. It is considered one of the most valuable plants in the tropics and is one of the world's chief sources of vegetable fat. The coconut will germinate even after floating in the ocean for 4 months.   Zone 10

352   *Copernicia hospita*

353   *Corypha umbraculifera*

353A   *Corypha umbraculifera*

354   *Cycas circinalis*

Copernicia hospita   Cuban Wax Palm
Family: Palmae
Origin: Central Cuba

352   One of the small group of wax palms. This one, native to Cuba, has a small, dense head, to 25 ft. high, and waxy, 3 ft. long leaves.   Zone 10

Corypha umbraculifera   Talipot Palm
Family: Palmae
Origin: India; Sri Lanka

353   The trunk of this tree grows to 80 ft. The leaves have a 16 ft. spread, as does the flower, making it the largest flower in the plant kingdom. It reaches maturity in 50–60 years, when it flowers and the leaves dry and drop off. An enormous blooming spadix then emerges from the top. Almost a year later, when the seeds have matured (1–2 tons), the top bends over and the tree falls.   Zone 10
    Pictures taken at Fairchild Gardens, Miami, Florida.

Cycas circinalis   Fern Palm
                                                Queen Cycas
Family: Cycadaceae   Sago Palm
Origin: East Africa; East Indies

354   This fast-growing cycad, which is not a true palm but resembles one, will grow to 20 ft. The stiff, glossy, dark green leaves grow up to 8 ft.   Zone 10
    Picture from Fairchild Gardens, Miami, Florida.

355   Cycas revoluta

355A   Cycas revoluta

356   Dioon edule

357   Dioon spinulosum

357A   Dioon spinulosum

Cycas revoluta   Japanese Fern Palm
                                                                                  Sago Palm
Family: Cycadaceae
Origin: Japan

355   The most rugged of all of the cycads and the most picturesque, it will grow to 10 ft. in full sun but in the desert it does
best with some shade. It is generally very slow growing, with shiny,
deep green, 3–4 ft. long foliage, and is an excellent tub
plant.   Zone 9

Dioon edule   Chestnut Dioon
Family: Zamiaceae
Origin: Central America

356   This palm-like plant is very stocky, growing to only about 6
ft. The sharp, pointed leaves are up to 6 ft. long. It requires partial
shade with good moisture. The seeds are edible.   Zone 10

Dioon spinulosum   Chamal
                                                                             Giant Dioon
Family: Zamiaceae
Origin: Southern Mexico

357   A cycad, to 20 ft., with a thick trunk. There is a dense, palm-
like rosette of leathery, 4–6 ft. long leaves with sharply pointed
leaflets. There are both male and female plants: the male cone is 1
ft. or longer; the female cone is low and rounded, with many
seeds.   Zone 10

358   Hyophorbe lagenicaulis

360   Latania loddigesii

359   Jubaea chilensis

361   Licuala grandis

361A   Licuala grandis

Hyophorbe lagenicaulis   **Bottle Palm**
Family: Palmae
Origin: Mascarene Islands
358   This interesting feather palm is swollen at the base and narrows upward, like a bottle. The trunk is gray and smooth up to the 2 ft. crownshaft, which is bright green. It is widely planted in the tropics and will not stand any frost.   Zone 10

Jubaea chilensis (J. spectabilis)   **Chilean Wine Palm**
Family: Palmae   **Coquito Palm**
Origin: Central coast of Chile
359   A tall feather palm to 60 ft. with a massive trunk, sometimes more than 3 ft. in diameter. The sap is called "palm honey," but, unfortunately, the tree has to be cut down to obtain it. It is now an endangered species and is protected by law in Chile.   Zone 10

Latania loddigesii   **Blue Latan Palm**
Family: Palmae
Origin: Mauritius Island
360   A blue palm that grows to 50 ft. in its native habitat but in cultivation grows slowly to only about 20 ft. The leaf blades are blue and glaucous, up to 5 ft. long and as wide.   Zone 10

Licuala grandis
Family: Palmae
Origin: New Hebrides
361   A palm tree that grows to about 10 ft. The handsome, dark green leaves are up to 3 ft. across on 3 ft. stems and have a very evenly "cut" outer edge.   Zone 10

Livistona chinensis   Chinese Fan Palm
Family: Palmae
Origin: China; Malaysia
362   A slow-growing, robust fan palm to 50 ft. It has a grayish-brown trunk, the crown is dense, with leaf stalks shorter than the leaf blades. It is a good indoor container plant.   Zone 10

363   *Livistona rotundifolia*

362   *Livistona chinensis*

Livistona rotundifolia (L. altissima)   Sadang
Family: Palmae
Origin: Philippines to Indonesia
363   A tall, very slender palm to 100 ft., with a brownish-gray trunk, slightly pendulous leaves, and a fairly dense crown. The interesting fruit is bright scarlet when young, turning black when ripe.   Zone 10

Neodypsis decaryi   Triangle Palm
Family: Palmae
Origin: Madagascar
364   The unusual arrangement of the 15 ft. long leaves coming out of the trunk gives this plant an eye-catching appearance. The leaves account for its common name, "Triangle Palm." It will grow to 30 ft.   Zone 10

365   *Phoenix canariensis*

364   *Neodypsis decaryi*

Phoenix canariensis   Canary Island Palm
Family: Palmae
Origin: Canary Islands
365   A stout palm to 50 ft. or more with a trunk up to 3 ft. in diameter and a 40 ft. spread. It is too big for any but the large garden.   Zone 8

366   Phoenix reclinata

Phoenix reclinata (P. pomila)   African Wild Date Palm
                                                              Senegal Date Palm
Family: Palmae   Wild Date Palm
Origin: Tropical and southern Africa
366   A slender, clustering palm to 20 ft. or more with long, graceful, 10 ft. fronds. This plant branches out at the base when young so
it is not unusual to see 20 or more trunks in a clump.   Zone 10

367   Phoenix roebelenii

Phoenix roebelenii   Pygmy Date Palm
                                                             Roebelen Date Palm
Family: Palmae
Origin: Laos
367   A slender, graceful palm rarely more than 8 ft. with graceful,
fern-like leaves on a dense crown. It is widely grown everywhere
in the tropics and widely cultivated as an elegant house plant
in colder areas, where it needs moisture and good light
indoors.   Zone 10

Pigafetta filaris   Wanga Palm
Family: Palmae
Origin: Celebes
368   This large, fast-growing, tropical feather palm from open or
heavy rainfall areas can grow to 150 ft. It will not tolerate dry or
cold climates. The leaves are very thorny. In its native habitat it is
valued for its timber.   Zone 10

368   Pigafetta filaris

369   Pritchardia pacifica

Pritchardia pacifica   Fiji Fan Palm
Family: Palmae
Origin: Samoa; Fiji Islands
369   This form grows to 30 ft. with a trunk seldom more than 1 ft.
thick. The stiff, light green leaves are fan-like, about 3 ft. wide and 4
ft. long.   Zone 10

370   *Ptychosperma macarthurii*

371   *Rhapis excelsa*

372   *Roystonea regia*

373   *Sabal palmetto*

Ptychosperma macarthurii (P. hospitum;   Hurricane Palm
Kentia macarthurii)   Macarthur Palm
Family: Palmae
Origin: New Guinea
370   A slender, clump-forming feather palm to 20 ft. It is well
suited for container culture.   Zone 10

Rhapis excelsa (R. flabelliformis)   Bamboo Palm
Family: Palmae   Lady Palm
Origin: China; Japan
371   A beautiful, dense, 5–10 ft., clump-forming palm that
spreads from underground shoots, the clump becoming wider and
wider in the same manner as bamboo. In colder areas, it is used as a
container plant, both indoors and outdoors.   Zone 10

Roystonea regia (R. jenmanii;   Cuban Royal Palm
Oreodoxa regia)
Family: Palmae
Origin: Cuba
372   The grayish-green trunk of this outstanding, 75 ft. palm has
leaves about 12 ft. long, with the top of the trunk a shiny, bright
green. It is widely used in landscaping, but, unfortunately, has a
habit of dropping old leaves, which weigh up to 40 pounds
each.   Zone 10

Sabal palmetto (S. viatoris)   Cabbage Palm
Family: Palmae   Palmetto Palm
Origin: Southeastern United States
373   A robust fan palm to 80 ft. with a dark trunk and a dense,
compact crown of 4–8 ft. leaves. It will grow in almost any type of
soil—rich, damp, sandy, or poor.   Zone 9

375   *Trachycarpus fortunei*

74   *Syagrus coronata*

76   *Veitchia merrillii*

377   *Washingtonia filifera*

Syagrus coronata (S. quinquerfaria)	Licuri Palm
	Triangle Palm

Family: Palmae
Origin: Arid regions of Brazil

374   This tree grows to 30 ft. Its leaves are 8 ft. or more in length and emerge one above the other, giving a three-sided effect to the trunk.   Zone 10

Trachycarpus fortunei (Chamaerops excelsa; C. fortunei)	Chusan Palm
	Windmill Palm

Family: Palmae
Origin: Eastern Asia

375   This palm has been planted in the cooler areas. It is a very handsome tree to 40 ft. The trunk is densely covered with black, hairy fibers. The crown is a mat of small, slender, fan-palm leaves.   Zone 8

Veitchia merrillii (Adonidia merrillii)	Manila Palm

Family: Palmae
Origin: Philippines

376   A slender palm to 15 ft. with a compact crown comprised of 10–12 bright green, arching leaves about 6 ft. long.   Zone 10

Washingtonia filifera (W. filamentosa)	California Fan Palm
	Desert Fan Palm

Family: Palmae
Origin: Along streams and springs of southern California and southwestern Arizona

377   This 50 ft., fast-growing, drought-resistant palm has large, 4–5 ft., broad, fan-shaped leaves. The plant does not clean itself so the dried leaves become a fire hazard if not cut off.   Zone 9

378   *Washingtonia robusta*

Washingtonia robusta (W. gracilis;   Mexican Fan Palm
W. sonorae; Pritchardia robusta)   Thread Palm

Family: Palmae
Origin: Mexico, from Baja California to southern Sonora

378   This palm has a tall, slender trunk, often up to 100 ft., with an 8 ft. diameter head at the top. It is widely used in southern California as a picturesque street tree.   Zone 9

COMMON/BOTANIC NAMES

COMMON	BOTANIC	COMMON	BOTANIC
Achiote	*Bixa orellana*	Barbados Aloe	*Aloe barbadensis*
African Evergreen	*Syngonium podophyllum*	Barbados Flower Fence	*Caesalpinia pulcherrima*
African Iris	*Dietes bicolor*	Barbados Lily	*Hippeastrum vittatum*
African Iris	*Dietes iridioides*	Barometer Bush	*Leucophyllum frutescens*
African Mallow	*Dombeya wallichii*	Beach Morning Glory	*Ipomoea pes-caprae*
African Nutmeg	*Monodora myristica*	Beach Naupaka	*Scaevola sericia*
African Tulip Tree	*Spathodea campanulata*	Bear's Breech	*Acanthus mollis*
African Tulip Tree	*Spathodea campunulata aurea*	Beefwood	*Casuarina equisetifolia*
		Be-still Tree	*Thevetia peruviana*
African Wild Date Palm	*Phoenix reclinata*	Bidwill Coral Tree	*Erythrina × bidwillii*
Alexandra Palm	*Archontophoenix alexandrae*	Bird of Paradise	*Strelitzia reginae*
		Bird's Eye Bush	*Ochna kirkii*
Allspice	*Pimenta dioica*	Black-eyed Susan Vine	*Thunbergia alata*
Amaryllis	*Hippeastrum vittatum*	Bleeding Heart Vine	*Clerodendrum thomsonae*
Amazon Water Lily	*Victoria amazonica*	Blood Trumpet Vine	*Distictis buccinatoria*
Anacahutta	*Cordia boissiere*	Blue Bird Vine	*Petrea volubilis*
Angels Trumpet	*Brugmansia suaveolens*	Blue Ginger	*Dichorisandra thyrsiflora*
Angel Wing Jasmine	*Jasminum nitidum*	Blue Latan Palm	*Latania loddigesii*
Annatto	*Bixa orellana*	Blue Potato Vine	*Solanum rantonnetii*
Ape	*Alocasia macrorrhiza*	Blue Sage	*Eranthemum pulchellum*
Apple Blossom	*Cassia javanica*	Bottlebrush	*Callistemon citrinus*
Apricot Moonflower	*Brugmansia vericolor*	Bottle Palm	*Hyophorbe lagenicaulis*
Arabian Coffee	*Coffea arabica*	Bottle Ponytail	*Beaucarnea recurvata*
Areca Palm	*Chrysalidocarpus lutescens*	Bottle Tree	*Brachychiton populneus*
		Bower Plant	*Pandorea jasminoides*
Arrowhead Vine	*Syngonium podophyllum*	Brazilian Flame Bush	*Calliandra tweedii*
Ashanti Blood	*Mussaenda erythrophylla*	Brazilian Golden Vine	*Stigmaphyllon ciliatum*
Asoka Tree	*Saraca indica*	Brazilian Nightshade	*Solanum seaforthianum*
Australian Flame Tree	*Brachychiton acerifolius*	Brazilian Pepper Tree	*Schinus terebinthifolius*
Australian Pine	*Casuarina equisetifolia*	Brazilian Plume Flower	*Justicia carnea*
Australian Tree Fern	*Alsophila australis*	Brazilian Red Cloak	*Megaskepasma erythrochlamys*
Autograph Tree	*Clusia rosea*		
Baby Sun Rose	*Aptenia cordifolia*	Breadfruit	*Artocarpus altilis*
Bagnit Vine	*Tristellateia australasiae*	Bridal Bouquet	*Stephanotis floribunda*
Bald Cypress	*Taxodium distichum*	Brisbane Box	*Tristania conferta*
Bamboo Palm	*Rhapis excelsa*	Bronze Dracaena	*Cordyline australis atropurpurea*
Banyan Tree	*Ficus benghalensis*		
Baobab	*Adansonia digitata*	Bronze Euphorbia	*Euphorbia cotinifolia*

COMMON	BOTANIC	COMMON	BOTANIC
Brownea	*Brownea capitella*	Climbing Lily	*Gloriosa rothschildiana*
Buddha's Lamp	*Mussaenda philippica* 'Donna Aurora'	Clustered Fishtail Palm	*Caryota mitis*
		Cockspur Coral Tree	*Erythrina crista-galli*
Buddhist Bauhinia	*Bauhinia variegata* 'Candida'	Coconut Palm	*Cocos nucifera*
		Common Gardenia	*Gardenia jasminoides*
Bugleweed	*Ajuga reptans*	Common Lantana	*Lantana camara*
Bulrush	*Cyperus papyrus*	Common Morning Glory	*Ipomoea purpurea*
Bunya-Bunya	*Araucaria bidwillii*	Common Screw Pine	*Pandanus utilis*
Burmese Fan Palm	*Caryota mitis*	Copper-leaf	*Acalypha wilkesiana*
Burning Bush	*Combretum microphyllum*	Coquito Palm	*Jubaea chilensis*
		Coral Plant	*Jatropha multifida*
Burning Love	*Ixora coccinea*	Coral Plant	*Russelia equisetifolia*
Bush Allamanda	*Allamanda neriifolia*	Coral Tree	*Erythrina caffra*
Bush Clockvine	*Thunbergia erecta*	Coral Vine	*Antigonon leptopus*
Buttercup Tree	*Cochlospermum vitifolium*	Corn Plant	*Dracaena fragrans* 'Massangeana'
Butterfly Flower	*Bauhinia monandra*	Costa Rican Nightshade	*Solanum wendlandii*
Butterfly Iris	*Dietes iridioides*	Crane Flower	*Strelitzia reginae*
Butterfly Palm	*Chrysalidocarpus lutescens*	Crape Ginger	*Costus speciosus*
		Crape Jasmine	*Stemmadenia galeottiana*
Butterfly Vine	*Stigmaphyllon ciliatum*	Crape Myrtle	*Lagerstroemia indica*
Cabbage Palm	*Sabal palmetto*	Croton	*Codiaeum variegatum*
Caimito	*Chrysophyllum cainito*	Crown Flower	*Calotropis gigantea*
Cajeput Tree	*Melaleuca quinquenervia*	Crown of Gold	*Cassia excelsa*
Calico Flower	*Aristolochia elegans*	Crown of Thorns	*Euphorbia milii*
California Fan Palm	*Washingtonia filifera*	Cry-Baby Tree	*Erythrina crista-galli*
Canarybird Bush	*Crotalaria agatiflora*	Cuban Royal Palm	*Roystonea regia*
Canary Island Date Palm	*Phoenix canariensis*	Cuban Wax Palm	*Copernicia hospita*
Candebury Tree	*Aleurites moluccana*	Cup of Gold	*Solandra maxima*
Candelabra Plant	*Aloe arborescens*	Dagger Plant	*Yucca aloifolia*
Candlenut Tree	*Aleurites moluccana*	Day Lily	*Hemerocallis lilioasphodelus*
Candlestick Senna	*Cassia alata*		
Candy Corn Vine	*Wagatea spicata*	Dead-rat Tree	*Adansonia digitata*
Cannonball Tree	*Couroupita guianensis*	Desert Fan Palm	*Washingtonia filifera*
Cape Chestnut	*Calodendrum capense*	Desert Willow	*Pittosporum phillyraeoides*
Cape Honeysuckle	*Tecomaria capensis*		
Cape Jasmine	*Gardenia jasminoides*	Divorce Vine	*Solanum wendlandii*
Cape Plumbago	*Plumbago auriculata*	Dracaena Palm	*Cordyline indivisa*
Cardinal's Guard	*Odontonema strictum*	Dragon Tree	*Dracaena draco*
Caricature Plant	*Graptophyllum pictum*	Dumb Cane	*Dieffenbachia* ssp.
Carrotwood	*Cupaniopsis anacardioides*	Dwarf Poinciana	*Caesalpinia pulcherrima*
		Easter-Lily Vine	*Beaumontia grandiflora*
Cassia	*Cassia excelsa*	East India Fig	*Ficus benghalensis*
Centipede Plant	*Homalocladium platycladum*	Edible Banana Plantain	*Musa* × *paradisiaca*
		Egyptian Paper Plant	*Cyperus papyrus*
Chain of Love Vine	*Antigonon leptopus*	Elderberry	*Sambucus mexicana* var. *bipinnata*
Chamal	*Dioon spinulosum*		
Chenille Plant	*Acalypha hispida*	Eldorado	*Pseuderanthemum reticulatum*
Chestnut Dioon	*Dioon edule*		
Chicle Tree	*Manilkara zapota*	Elephant Foot Tree	*Beaucarnea recurvata*
Chilean Pine	*Araucaria araucana*	Elephant's Ear	*Colocasia esculenta*
Chilean Wine Palm	*Jubaea chilensis*	Elphin Herb	*Cuphea hyssopifolia*
Chinaberry Tree	*Melia azedarach*	False Heather	*Cuphea hyssopifolia*
Chinese Box	*Murraya paniculata*	False Hop	*Justicia brandegeana*
Chinese Fan Palm	*Livistona chinensis*	False Kamani	*Terminalia catappa*
Chinese Hat Plant	*Holmskioldia sanguinea*	Fern Palm	*Cycas circinalis*
Christmas Berry Tree	*Schinus terebinthifolius*	Fern Pine	*Podocarpus gracilior*
Christmas Candle	*Cassia alata*	Fern Tree	*Filicium decipiens*
Christmas Flower	*Euphorbia pulcherrima*	Fiddleleaf Fig	*Ficus lyrata*
Christ Thorn	*Euphorbia milii*	Fiji Fan Palm	*Pritchardia pacifica*
Chusan Palm	*Trachycarpus fortunei*	Finger Tree	*Euphorbia tirucalli*
Cigar Flower	*Cuphea ignea*	Firebird	*Heliconia bihai* var. *aurea*

COMMON	BOTANIC	COMMON	BOTANIC
Firebush Bush	Streptosolen jamesonii	Heavenly Bamboo	Nandina domestica
Firecracker Plant	Cuphea ignea	Herald's Trumpet	Beaumontia grandiflora
Fire Cracker Plant	Russelia equisetiformis	Hidden Lily	Curcuma cocana
Fire-on-the-Mountain	Euphorbia cotinifolia	Honey Plant	Hoya carnosa
Fire Tree	Spathodea campanulata	Hong Kong Orchid Tree	Bauhinia blakeana
Firewheel Tree	Oreocallis pinnata	Horsetail Tree	Casuarina equisetifolia
Firewheel Tree	Stenocarpus sinuatus	Huamuchil	Pithecellobium dulce
Fishtail Palm	Caryota cumingii	Hurricane Palm	Ptychosperma
Fishtail Palm	Caryota no		macarthurii
Flamboyant	Delonix regia	Hydrangea Tree	Dombeya wallichii
Flame Bottle Tree	Brachychiton acerifolius	Indian Coral Tree	Erythrina variegata
Flame Creeper	Combretum	Indian Laburnum	Cassia fistula
	microphyllum	Indian Laurel Fig	Ficus retusa var. nitida
Flame-of-the-Forest	Butea monosperma	India Rubber Plant	Ficus elastica
Flame of the Forest	Spathodea campanulata	Ixora	Ixora odorata
Flame Tree	Brachychiton acerifolius	Jacaranda	Jacaranda mimosifolia
Flame Tree	Delonix regia	Jacob's Coat	Acalypha wilkesiana
Flame Vine	Pyrostegia venusta	Jacob's Coat	Acalypha wilkesiana
Flamingo Lily	Anthurium andraeanum		'Godseffiana'
Floradora	Stephanotis floribunda	Japanese Aralia	Fatsia japonica
Floss Silk Tree	Chorisia speciosa	Japanese Fern Palm	Cycas revoluta
Fountain Bush	Russelia equisetiformis	Japanese Hibicus	Hibiscus schizopetalus
Fountain Grass	Pennisetum setaceum	Japanese Lantern	Hibiscus schizopetalus
Fountain Tree	Spathodea campanulata	Japanese Medlar	Eriobotrya japonica
Frangipani	Plumeria rubra	Japanese Plum	Eriobotrya japonica
Frangipani Tree	Plumeria obtusa	Japanese Yew Pine	Podocarpus macrophyllus
	'Singapore'	Java Glorybower	Clerodendrum
Fruta Bomba	Carica papaya		speciosissimum
Garland Flower	Hedychium coronarium	Jelly Palm	Butia capitata
Garlic Vine	Pseudocalymma alliaceum	Jerusalem Thorn	Parkinsonia aculeata
Geiger Tree	Cordia sebestena	Joy Weed	Alternanthera ficoidea
Giant Bird of Paradise	Strelitzia nicolai		'Versicolor'
Giant Burmese	Lonicera hildebrandiana	Jungle Flame	Ixora coccinea
Honeysuckle		Jungle Geranium	Ixora coccinea
Giant Dioon	Dioon spinulosum	Kaffirboom	Erythrina caffra
Giant Elephant's Ear	Alocasia macrorrhiza	Kaffir Lily	Clivia miniata
Giant Ixora	Ixora duffii	Kaffir Plum	Harpephyllum caffrum
Giant Spiral Ginger	Tapeinochilus ananassae	Kahili Ginger	Hedychium gardnerianum
Giant Timber Bamboo	Bambusa oldhamii	Kalo	Colocasia esculenta
Glory Bower	Clerodendrum thomsonae	Kapok	Ceiba pentandra
Glory Bush	Tibouchina urvilleana	King Palm	Archontophoenix
Gold Creeper	Stigmaphyllon		cunninghamiana
	emarginatum	King's Mantle	Thunbergia erecta
Golden Bell	Tabebuia caraiba	Kino	Coccoloba uvifera
Golden Candle	Pachystachys lutea	Kolomona	Cassia surattensis
Golden Dewdrop	Duranta repens	Kukui	Aleurites moluccana
Golden Philodendron	Epipremnum aureum	Lady-of-the-Night	Brunfelsia nitida
Golden Pothos	Epipremnum aureum	Lady Palm	Rhapis excelsa
Golden Shower	Cassia fistula	Laua'e	Microsorium scolopendria
Golden Trumpet	Allamanda cathartica	Laurel-leaved Snailseed	Cocculus laurifolius
Gold Tree	Tabebuia donnell-smithii	Lavender Trumpet Vine	Clytostoma callistegioides
Good Luck Plant	Coryline terminalis	Leadwort	Plumbago auriculata
Grant Potato Vine	Solanum wendlandii	Lecheso	Stemmadenia galeottiana
Gray Goddess	Brahea armata	Leechee	Litchi chinensis
Green Jade Vine	Strongylodon macrobotrys	Lemon Lily	Hemerocallis
Guadalupe Fan Palm	Brahea edulis		lilioasphodelus
Guiana Chestnut	Pachira aquatica	Licuri Palm	Syagrus coronata
Hair Palm	Chamaerops humilis	Lipstick Tree	Bixa orellana
Hapu'u-ii	Cibotium chamissoi	Litchi	Litchi chinensis
Hardy Blue Cocos	Butia capitata	Little-leaf Fig	Ficus rubiginosa
Hau Tree	Hibiscus tiliaceus	Lobster Claw	Heliconia humilis
Hawaiian Tree Fern	Cibotium chamissoi	Loquat	Eriobotrya japonica

COMMON	BOTANIC	COMMON	BOTANIC
Love-charm	*Clytostoma callistegioides*	Norfolk Island Pine	*Araucaria heterophylla*
Lucky Nut	*Thevetia peruviana*	Octopus Tree	*Brassaia actinophylla*
Lychee	*Litchi chinensis*	Ohi'a-Lehua	*Metrosideros collina*
Macarthur Palm	*Ptychosperma macarthurii*	Oleander	*Nerium oleander*
		Opiuma	*Pithecellobium dulce*
Macaw Flower	*Heliconia bihai* var. *aurea*	Orange Glow Vine	*Senecio confusus*
Madagascar Jasmine	*Stephanotis floribunda*	Orange Jessamine	*Murraya paniculata*
Madre	*Gliricidia sepium*	Orchid Tree	*Bauhinia variegata* 'Candida'
Mahoe Tree	*Hibiscus tiliaceus*		
Maikoa	*Brugmansia suaveolens*	Our Lord's Candle	*Yucca whipplei*
Maile-scented Fern	*Microsorium scolopendria*	Pagoda Tree	*Plumeria obtusa* 'Singapore'
Malay Apple	*Syzygium malaccense*		
Malay Ginger	*Costus speciosus*	Palmetto Palm	*Sabal palmetto*
Malay Ixora	*Ixora duffii*	Palm Lily	*Cordyline stricta*
Malaysian Orchid	*Medinilla magnifica*	Pandang Screw Pine	*Pandanus odoratissimus*
Mallet Flower	*Tupidanthus calyptratus*	Papaya	*Carica papaya*
Mallow Rose	*Hibiscus moscheutos*	Paperbark Tree	*Melaleuca quinquenervia*
Mandarin's Hat	*Holmskioldia sanguinea*	Papyrus	*Cyperus papyrus*
Mango	*Mangifera indica*	Paraguay Nightshade	*Solanum rantonnettii*
Manila Palm	*Veitchia merrillii*	Parrot's Flower	*Heliconia psittacorum*
Marmalade Bush	*Streptosolen jamesonii*	Parrot's Plantain	*Heliconia psittacorum*
Marriage Vine	*Solanum wendlandii*	Parrot Tree	*Butea monosperma*
Match-Me-If-You-Can	*Acalypha wilkesiana*	Pascuita	*Euphorbia leucocephala*
Mauna Loa	*Spathiphyllum* 'Clevelandii'	Passion Flower	*Passiflora* × *alatocaerulea*
		Passion Fruit	*Passiflora edulis*
Medicinal Aloe	*Aloe barbadensis*	Pawpaw	*Carica papaya*
Mediterranean Fan Palm	*Chamaerops humilis*	Pelican Flower	*Aristolochia gigantea*
Mexican Blue Palm	*Brahea armata*	Pencil Tree	*Euphorbia tirucalli*
Mexican Creeper	*Antigonon leptopus*	Peregrina	*Jatropha integerrima*
Mexican Fan Palm	*Washingtonia robusta*	Persian Lilac	*Melia azedarach*
Mexican Flame Bush	*Calliandra tweedii*	Phanera	*Bauhinia corymbosa*
Mexican Flameleaf	*Euphorbia pulcherrima*	Philippine Wax Flower	*Nicolaia elatior*
Mexican Flame Vine	*Senecio confusus*	Physic Nut	*Jatropha multifida*
Mexican Gold Bush	*Galphimia glauca*	Piccabeen Palm	*Archontophoenix cunninghamiana*
Mexican Palo Verde	*Parkinsonia aculeata*		
Mickey Mouse Plant	*Ochna kirkii*	Pigeon Berry	*Duranta repens*
Milkbush	*Euphorbia tirucalli*	Pimento	*Pimenta dioica*
Mimosa Tree	*Albizia julibrissin*	Pincushion Flower	*Leucospermum*
Mindanao Gum	*Eucalyptus deglupta*	Pineapple Guava	*Feijoa sellowiana*
Mock Orange	*Pittosporum undulatum*	Pink Acacia	*Albizia julibrissin*
Monkey-bread Tree	*Adansonia digitata*	Pink-and-White Shower Tree	*Cassia javanica*
Monkey Pod Tree	*Samanea saman*		
Monkey Puzzle Tree	*Araucaria araucana*	Pink Jasmine	*Jasminum polyanthum*
Montezuma Cypress	*Taxodium mucronatum*	Pink Mandevilla	*Mandevilla splendens*
Moses-in-a-Boat	*Rhoeo spathacea*	Pink Orchid Tree	*Bauhinia monandra*
Moses-in-a-Cradle	*Rhoeo spathacea*	Pink Passion Flower	*Passiflora jamesonii*
Mother of Chocolate	*Gliricidia sepium*	Pink Porcelain Lily	*Alpinia zerumbet*
Mountain Ebony	*Bauhinia variegata* 'Candida'	Pink Poui-Rosea	*Tabebuia rosea*
		Pink Powderpuff	*Calliandra haematocephala*
Mudar	*Calotropis gigantea*		
Mysore Trumpet Vine	*Thunbergia mysorensis*	Pink Shower Orchid	*Congea tomentosa*
Naked Coral Tree	*Erythrina coralloides*	Pink Trumpet Tree	*Tabebuia rosea*
Narrow-leafed Pittosporum	*Pittosporum phillyraeoides*	Pitanga	*Eugenia uniflora*
		Platterleaf	*Coccoloba uvifera*
Naseberry	*Manilkara zapota*	Poinsettia	*Euphorbia pulcherrima*
Natal Coral Tree	*Erythrina humeana*	Poison Bulb Lily	*Crinum asiaticum*
Natal Plum	*Carissa macrocarpa*	Polygala	*Securidaca diversifolia*
Nephthytis	*Syngonium podophyllum*	Pomeral Jambos	*Syzygium malaccense*
New Guinea Creeper	*Mucuna bennettii*	Port Jackson Fig	*Ficus rubiginosa*
Nicaraguan Cocoa-Shade	*Gliricidia sepium*	Pride of Barbados	*Caesalpinia pulcherrima*
Night-blooming Cereus	*Hylocereus undatus*	"Pride of India"	*Lagerstroemia speciosa*
Niroli	*Filicium decipiens*	Pride of India	*Melia azedarach*

COMMON	BOTANIC	COMMON	BOTANIC
Pride of Madeira	*Echium fastuosum*	Satinwood	*Murraya paniculata*
Primavera	*Tabebuia donnell-smithii*	Sausage Tree	*Kigelia pinnata*
Primrose Jasmine	*Jasminum mesnyi*	Scottish Attorney	*Clusia rosea*
Prince Kuhio Vine	*Ipomoea horsfalliae*	Sea Grape	*Coccoloba uvifera*
Princess Flower	*Tibouchina urvilleana*	Senegal Date Palm	*Phoenix reclinata*
Provision Tree	*Pachira aquatica*	Shampoo Ginger	*Costus spicatus*
Pua-kenikeni	*Fagraea berteriana*	Shaving Brush Tree	*Pseudobombax ellipticum*
Purple Allamanda	*Allamanda violacea*	Shell Ginger	*Alpinia zerumbet*
Purple Granadilla	*Passiflora edulis*	Shower of Gold	*Galphimia glauca*
Pygmy Date Palm	*Phoenix roebelenii*	Shower-of-Orchids	*Congea tomentosa*
Queen Cycas	*Cycas circinalis*	Shrimp Plant	*Justicia brandegeana*
Queen Emma Lily	*Crinum augustum* 'Queen Emma'	Shrub Verbena	*Lantana camara*
		Silk-Cotton Tree	*Ceiba pentandra*
Queen of the Night	*Hylocereus undatus*	Silk Oak	*Grevillea robusta*
Queen Palm	*Arecastrum romanzoffianum*	Silk Tree	*Albizia julibrissin*
		Silver Palm	*Coccothrinax alta*
Queen's Crape Myrtle	*Lagerstroemia speciosa*	Silver Tree	*Leucadendron argenteum*
Queensland Nut	*Macadamia integrifolia*	Silver Tree	*Tabebuia caraiba*
Queensland Pittosporum	*Pittosporum rhombifolium*	Simal	*Bombax ceiba*
		Sky Flower	*Duranta repens*
Queensland Umbrella Tree	*Brassaia actinophylla*	Sky Flower	*Thunbergia grandiflora*
		Sleeping Hibiscus	*Malvaviscus arboreus*
Queen's Wreath	*Petrea volubilis*	Small-leafed Strelitzia	*Strelitzia reginae* var. *juncea*
Railroad Vine	*Ipomoea pes-caprae*		
Rainbow Shower Tree	*Cassia javanica* × *C. fistula*	Small-leaf Rubber Tree	*Ficus benjamina*
		Smooth-Shelled Macadamia	*Macadamia integrifolia*
Rain Tree	*Samanea saman*		
Rattlesnake Plant	*Calathea insignis*	Snow Bush	*Breynia disticha* 'Roseo-picta'
Red Abyssinian Banana	*Ensete ventricosum* ''Maurellii''		
		Society Garlic	*Tulbaghia violacea*
Red Banana	*Ensete ventricosum* 'Maurellii'	Soft-tip Yucca	*Yucca gloriosa*
		Sorrowless Tree	*Saraca indica*
Red Bauhinia	*Bauhinia galpinii*	Spanish Bayonet	*Yucca aloifolia*
Red Cats-tail	*Acalypha hispida*	Spanish Dagger	*Yucca gloriosa*
Red Flag Bush	*Mussaenda erythrophylla*	Spathe Flower	*Spathiphyllum* 'Clevelandii'
Red Flame	*Hemigraphis alternata*		
Red Ginger	*Alpinia purpurata*	Spider Lily	*Crinum asiaticum*
Red Granadilla	*Passiflora coccinea*	Split-leaf Philodendron	*Monstera deliciosa*
Red Ivy	*Hemigraphis alternata*	Squirrels Tail	*Justicia betonica*
Red Jade	*Mucuna bennettii*	Star Apple	*Chrysophyllum cainito*
Red Passion Flower	*Passiflora coccinea*	Star Jasmine	*Jasminum multiflorum*
Red Silk Cotton Tree	*Bombax ceiba*	Star Potato Vine	*Solanum seaforthianum*
Red Spurge	*Euphorbia cotinifolia*	Strawberry Guava	*Psidium cattleianum*
Rice-paper Plant	*Tetrapanax papyriferus*	Sugarcane	*Saccharum officinarum*
Roebelen Date Palm	*Phoenix roebelenii*	Surinam Cherry	*Eugenia uniflora*
Roman Candle	*Yucca gloriosa*	Swamp Cypress	*Taxodium distichum*
Rose Apple	*Syzygium malaccense*	Swamp Immortelle	*Erythrina fusca*
Rose Bay	*Nerium oleander*	Swamp Rose Mallow	*Hibiscus moscheutos*
Rothschild Glorylily	*Gloriosa rothschildiana*	Swamp Tea Tree	*Melaleuca quinquenervia*
Roxburg Fig	*Ficus auriculata*	Sweet Clockvine	*Thunbergia fragrans*
Royal Poinciana	*Delonix regia*	Sweetheart Vine	*Pyrostegia venusta*
Royal Water Lily	*Victoria amazonica*	Sweet Shade	*Hymenosporum flavum*
Rubber Vine	*Cryptostegia grandiflora*	Swiss Cheese Plant	*Monstera deliciosa*
Rusty Fig	*Ficus rubiginosa*	Tahinu	*Messerschmidia argentea*
Sacred Bamboo	*Nandina domestica*	Talipot Palm	*Corypha umbraculifera*
Sadang	*Livistona rotundifolia*	Tapeworm Bush	*Homalocladium platycladum*
Sago Palm	*Cycas circinalis*		
Sago Palm	*Cycas revoluta*	Taro	*Colocasia esculenta*
St. Thomas Tree	*Bauhinia monandra*	Tasmanian Tree Fern	*Dicksonia antarctica*
Sanchezia	*Sanchezia nobilis* var. *glaucophylla*	Teak	*Tectona grandis*
		Temple Tree	*Plumeria rubra*
Sapodilla	*Manilkara zapota*	Texas Ranger	*Leucophyllum frutescens*

COMMON	BOTANIC	COMMON	BOTANIC
Texas Silver Leaf	*Leucophyllum frutescens*	Wanga Palm	*Pigafetta filaris*
Thread Palm	*Washingtonia robusta*	Water-Platter	*Victoria amazonica*
Tiger's Claw	*Erythrina variegata*	Wax Mallow	*Malvaviscus arboreus*
Timber Bamboo	*Bambusa oldhamii*	Wax Plant	*Hoya carnosa*
Timor Screw Pine	*Pandanus baptistii aureus*	Weeping Bottlebush	*Callistemon viminalis*
Ti Plant	*Cordyline terminalis*	Weeping Fig	*Ficus benjamina*
Torch Ginger	*Nicolaia elatior*	Weeping Lantana	*Lantana montevidensis*
Torch Plant	*Aloe arborescens*	Wheel of Fire	*Stenocarpus sinuatus*
Trailing Lantana	*Lantana montevidensis*	White Ginger	*Hedychium coronarium*
Trailing Wedelia	*Wedelia trilobata*	White Lace Euphorbia	*Euphorbia leucocephala*
Travelers Palm	*Ravenala madagascariensis*	White Potato Vine	*Solanum jasminoides*
		White Shrimp Plant	*Justicia betonica*
Travelers Tree	*Ravenala madagascariensis*	Wild Cocoa Tree	*Pachira aquatica*
		Wild Cotton	*Cochlospermum vitifolium*
Tree Aloe	*Aloe arborescens*		
Tree Heliotrope	*Messerschmidia argentea*	Wild Cotton	*Hibiscus moscheutos*
Tree of Gold	*Tabebuia caraiba*	Wild Date Palm	*Phoenix reclinata*
Triangle Palm	*Neodypsis decaryi*	Wild Plantain	*Heliconia bihai* var. *aurea*
Triangle Palm	*Syagrus coronata*	Wild Poinsettia	*Warszewiczia coccinea*
Tropical Almond	*Terminalia catappa*	Windmill Jasmine	*Jasminum nitidum*
Trumpet Creeper	*Campsis* × *tagliabuana* 'Madame Galen'	Windmill Palm	*Trachycarpus fortunei*
		Wood Rose	*Ipomoea tuberosa*
Trumpet Plant	*Solandra maxima*	Yellow Allamanda	*Allamanda cathartica*
Trumpet Tree	*Tabebuia serratifolia*	Yellow Elder	*Tecoma stans*
Tuckeroo	*Cupaniopsis anacardioides*	Yellow Flax	*Reinwardtia indica*
		Yellow Morning Glory	*Ipomoea tuberosa*
Umbrella Plant	*Cyperus alternifolius*	Yellow Oleander	*Thevetia peruviana*
Victorian Box	*Pittosporum undulatum*	Yellow Poui	*Tabebuia serratifolia*
Virgin Tree	*Mussaenda philippica* 'Donna Aurora'	Yesterday-Today-and-Tomorrow	*Brunfelsia pauciflora* 'Floribunda'

SYNONYMS

Achras zapota see *Manilkara zapota*
Adenoropium hastatum see *Jatropha integerrima*
Adenoropium integerrimum see *Jatropha integerrima*
Adenoropium multifidium see *Jatropha multifida*
Adonidia merrillii see *Veitchia merrillii*
Alocasia indica see *Alocasia macrorrhiza*
Aloe vera see *Aloe barbadensis*
Alpinia speciosa see *Alpinia zerumbet*
Alsophila cooperi see *Alsophila australis*
Amaryllis vittatum see *Hippeastrum vittatum*
Aralia japonica see *Fatsia japonica*
Aralia papyriferus see *Tetrapanax papyriferus*
Aralia sieboldi see *Fatsia japonica*
Araucaria excelsa see *Araucaria heterophylla*
Araucaria imbricata see *Araucaria araucana*
Aristolochia grandiflora see *Aristolochia gigantea*
Artocarpus incisus see *Artocarpus altilis*
Bauhinia punctata see *Bauhinia galpinii*
Beloperone guttata see *Justicia brandegeana*
Bignonia caerulea see *Jacaranda mimosifolia*
Bignonia capensis see *Tecomaria capensis*
Bignonia cherere see *Distictis buccinatoria*
Bignonia jasminoides see *Pandorea jasminoides*
Bignonia speciosa see *Clytostoma callistegioides*
Bignonia venusta see *Pyrostegia venusta*
Bignonia violacea see *Clytostoma callistegioides*
Bombax ellipticum see *Pseudobombax ellipticum*
Bombax malabaricum see *Bombax ceiba*
Breynia nivosa see *Breynia disticha* 'Roseo-picta'
Brunfelsia calycina see *Brunfelsia pauciflora* 'Floribunda'
Brunfelsia grandiflora see *Brunfelsia pauciflora* 'Macrantha'
Butea frondosa see *Butea monosperma*
Caladium esculentum see *Colocasia esculenta*
Calliandra inaequilatera see *Calliandra haematocephala*
Callistemon lanceolatus see *Callistemon citrinus*
Carissa grandiflora see *Carissa macrocarpa*
Caryota alberti see *Caryota no*
Caryota furfuraceae see *Caryota mitis*
Caryota griffithii see *Caryota mitis*

Caryota rumphiana see *Caryota no*
Caryota sobolifera see *Caryota mitis*
Cassia glauca see *Cassia surattensis*
Catimbrum speciosum see *Alpinia zerumbet*
Ceiba casearia see *Ceiba pentandra*
Chamaerops excelsa see *Trachycarpus fortunei*
Chamaerops fortunei see *Trachycarpus fortunei*
Cibotium splendens see *Cibotium chamissoi*
Clerodendrum fallax see *Clerodendrum speciosissimum*
Cocos australis see *Butia capitata*
Cocos plumosa see *Arecastrum romanzoffianum*
Cocos romanzoffianum see *Arecastrum romanzoffianum*
Cordyline congesta see *Cordyline stricta*
Costus cylindricus see *Costus spicatus*
Crinum floridanum see *Crinum asiaticum*
Croton pictum see *Codiaeum variegatum*
Cupania anacardioides see *Cupaniopsis anacardioides*
Cyathea cooperi see *Alsophila australis*
Cybistax donnell-smithii see *Tabebuia donnell-smithii*
Datura mollis see *Brugmansia versicolor*
Datura suaveolens see *Brugmansia suaveolens*
Dipladenia splendens see *Mandevilla splendens*
Dracaena australis atropurpurea see *Cordyline australis atropurpurea*
Dracaena indivisa see *Cordyline indivisa*
Dracaena stricta see *Cordyline stricta*
Dracaena terminalis see *Cordyline terminalis*
Drejella guttata see *Justicia brandegeana*
Duranta ellisia see *Duranta repens*
Duranta plumieri see *Duranta repens*
Embothrium wickhamii see *Oreocallis pinnata*
Eranthemum nervosum see *Eranthemum pulchellum*
Eranthemum reticulatum see *Pseuderanthemum reticulatum*
Eriodendron anfractuosum see *Ceiba pentandra*
Erythea armata see *Brahea armata*
Erythea edulis see *Brahea edulis*
Erythea roezlii see *Brahea armata*
Erythrina constantiana see *Erythrina caffra*
Erythrina glauca see *Erythrina fusca*

Erythrina indica see *Erythrina variegata*
Erythrina laurifolia see *Erythrina crista-galli*
Eugenia malaccensis see *Syzygium malaccense*
Euphorbia bojeri see *Euphorbia milii*
Euphorbia scotana see *Euphorbia cotinifolia*
Fatsia papyrifera see *Tetrapanax papyriferus*
Ficus australis see *Ficus rubiginosa*
Ficus belgica see *Ficus elastica*
Ficus indica see *Ficus benghalensis*
Ficus microcarpa var. *nitida* see *Ficus retusa* var. *nitida*
Ficus pandurata see *Ficus lyrata*
Ficus roxburghii see *Ficus auriculata*
Galphimia nitida see *Galphimia glauca*
Gardenia florida see *Gardenia jasminoides*
Gardenia grandiflora see *Gardenia jasminoides*
Gliricidia maculata see *Gliricidia sepium*
Graptophyllum hortense see *Graptophyllum pictum*
Heliconia collinsiana see *Heliconia pendula*
Heliconia distans see *Heliconia bihai* var. *aurea*
Heliconia elongata see *Heliconia wagnerana*
Heliconia stricta see *Heliconia wagnerana*
Hemerocallis flava see *Hemerocallis lilioasphodelus*
Hemigraphis colorata see *Hemigraphis alternata*
Hesperoyucca whipplei see *Yucca whipplei*
Hibiscus abutiloides see *Hibiscus tiliaceus*
Imantophyllum miniatum see *Clivia miniata*
Inga dulcis see *Pithecellobium dulce*
Inga pulcherrima see *Calliandra tweedii*
Ipomoea bibol see *Ipomoea pes-caprae*
Ipomoea briggsii see *Ipomoea horsfalliae*
Ipomoea mexicana see *Ipomoea purpurea*
Ixora bandhuca see *Ixora coccinea*
Ixora incarnata see *Ixora coccinea*
Ixora lutea see *Ixora coccinea*
Ixora macrothyrsa see *Ixora duffii*
Jacaranda ovalifolia see *Jacaranda mimosifolia*
Jacobinia carnea see *Justicia carnea*
Jambosa malaccensis see *Syzygium malaccense*
Jasminum magnificum see *Jasminum nitidum*
Jasminum primulinum see *Jasminum mesnyi*
Jasminum pubescens see *Jasminum multiflorum*
Jatropha hastata see *Jatropha integerrima*
Jubaea spectabilis see *Jubaea chilensis*
Kentia macarthurii see *Ptychosperma macarthurii*
Lagerstroemia elegans see *Lagerstroemia indica*
Lagerstroemia flos-reginae see *Lagerstroemia speciosa*
Languas speciosa see *Alpinia zerumbet*
Lantana delicata see *Lantana montevidensis*
Lantana delicatissima see *Lantana montevidensis*
Lantana sellowiana see *Lantana montevidensis*
Leucophyllum texanum see *Leucophyllum frutescens*
Linum trigynum see *Reinwardtia indica*
Livistona altissima see *Livistona rotundifolia*
Lycianthes cantonei see *Solanum rantonnettii*
Macadamia ternifolia see *Macadamia integrifolia*
Manilkara zapotella see *Manilkara zapota*
Medemia nobilis see *Bismarckia nobilis*
Melaleuca leucadendra see *Melaleuca quinquenervia*
Melia australis see *Melia azedarach*
Melia japonica see *Melia azedarach*
Melia sempervirens see *Melia azedarach*
Merremia tuberosa see *Ipomoea tuberosa*
Mesembryanthemum cordifolium see *Aptenia cordifolia*
Meyenia erecta see *Thunbergia erecta*

Moraea bicolor see *Dietes bicolor*
Moraea iridioides see *Dietes iridioides*
Muehlenbeckia platyclada see *Homalocladium platycladum*
Murraya exotica see *Murraya paniculata*
Musa acuminata × *M. balbisiana* see *Musa* × *paradisiaca*
Musa 'Maurellii' see *Ensete ventricosum* 'Maurellii'
Musa × *sapientum* see *Musa* × *paradisiaca*
Nephelium litchi see *Litchi chinensis*
Nephthytis afzelii see *Syngonium podophyllum*
Nerium indicum see *Nerium oleander*
Nerium odorum see *Nerium oleander*
Nicoteba betonica see *Justicia betonica*
Nolina tuberculata see *Beaucarnea recurvata*
Operculina tuberosa see *Ipomoea tuberosa*
Oreodoxa regia see *Roystonea regia*
Pachira fastuosa see *Pseudobombax ellipticum*
Pachira grandiflora see *Pachira aquatica*
Pachira macrocarpa see *Pachira aquatica*
Paritium liliaceum see *Hibiscus tiliaceus*
Passiflora alata × *P. caerulea* see *Passiflora* × *alatocaerulea*
Passiflora pfordtii see *Passiflora* × *alatocaerulea*
Pennisetum ruppelianum see *Pennisetum setaceum*
Phaedranthus buccinatorius see *Distictis buccinatoria*
Phaeomeria magnifica see *Nicolaia elatior*
Phaeomeria speciosa see *Nicolaia elatior*
Pharbitis purpurea see *Ipomoea purpurea*
Philodendron johnsii see *Philodendron selloum*
Philodendron pertusum see *Monstera deliciosa*
Phoenix pomila see *Phoenix reclinata*
Phyllathus nivosa see *Breynia disticha* 'Roseo-picta'
Pimenta officinalis see *Pimenta dioica*
Pithecellobium saman see *Samanea saman*
Pleroma grandiflora see *Tibouchina urvilleana*
Pleroma splendens see *Tibouchina urvilleana*
Plumbago capensis see *Plumbago auriculata*
Plumeria emarginata see *Plumeria obtusa* 'Singapore'
Podocarpus longifolius see *Podocarpus macrophyllus*
Poinciana pulcherrima see *Caesalpinia pulcherrima*
Poinciana regia see *Delonix regia*
Poinsettia pulcherrima see *Euphorbia pulcherrima*
Polygala diversifolia see *Securidaca diversifolia*
Polypodium phymatodes see *Microsorium scolopendria*
Polypodium scolopendria see *Microsorium scolopendria*
Pothos aureus see *Epipremnum aureum*
Pritchardia robusta see *Washingtonia robusta*
Psidium littorale see *Psidium cattleianum*
Ptychosperma alexandrae see *Archontophoenix alexandrae*
Ptychosperma cunninghamiana see *Archontophoenix cunninghamiana*
Ptychosperma hospitum see *Ptychosperma macarthurii*
Pyrostegia ignea see *Pyrostegia venusta*
Raphidophora aurea see *Epipremnum aureum*
Reinwardtia tetragyma see *Reinwardtia indica*
Reinwardtia trigynum see *Reinwardtia indica*
Rhapis flabelliformis see *Rhapis excelsa*
Rhoeo discolor see *Rhoeo spathacea*
Rhoeo tradescantia discolor see *Rhoeo spathacea*
Roystonea jenmanii see *Roystonea regia*
Russelia juncea see *Russelia equisetiformis*
Sabal viatoris see *Sabal palmetto*
Salmalia malabarica see *Bombax ceiba*
Saphota acras see *Manilkara zapota*

Scaevola frutescens see *Scaevola sericia*
Scaevola taccada see *Scaevola sericia*
Schefflera actinophylla see *Brassaia actinophylla*
Scindapsus aureus see *Epipremnum aureum*
Seaforthia elegans see *Archontophoenix cunninghamiana*
Sinocalamus oldhamii see *Bambusa oldhamii*
Solandra guttata see *Solandra maxima*
Solandra hartwegii see *Solandra maxima*
Solandra nitida see *Solandra maxima*
Sphaeropteris cooperi see *Alsophila australis*
Stenolobium stans see *Tecoma stans*
Sterculia acerifolia see *Brachychiton acerifolius*
Sterculia diversifolia see *Brachychiton populneus*
Strelitzia juncea see *Strelitzia reginae* var. *juncea*
Strelitzia parvifolia see *Strelitzia reginae*
Strelitzia parvifolia var. *juncea* see *Strelitzia reginae* var. *juncea*

Syagrus quinquerfaria see *Syagrus coronata*
Tabebuia pallida see *Tabebuia rosea*
Tabebuia pentaphylla see *Tabebuia rosea*
Tacsonia jamesonii see *Passiflora jamesonii*
Taxodium mexicanum see *Taxodium mucronatum*
Tecoma argentia see *Tabebuia caraiba*
Tecoma capensis see *Tecomaria capensis*
Tecoma jasminoides see *Pandorea jasminoides*
Thevetia neriifolia see *Thevetia peruviana*
Thryallis glauca see *Galphimia glauca*
Tibouchina grandiflora see *Tibouchina urvilleana*
Tibouchina semidecandra see *Tibouchina urvilleana*
Tournefortia argentea see *Messerschmidia argentea*
Victoria regia var. *randi* see *Victoria amazonica*
Washingtonia filamentosa see *Washingtonia filifera*
Washingtonia gracilis see *Washingtonia robusta*
Washingtonia sonorae see *Washingtonia robusta*

BIBLIOGRAPHY

Beard, J. S. *West Australian Plants.* Surrey Beatty & Sons, 1970. *Wildflowers of the Northwest: An Introduction to the Flora of Northwestern Australia.* Mercantile Press.

Blombery, Alec, and Rodd, Tony. *Palms.* Angus & Robertson Publishers, Australia, 1982.

Eliovson, Sima. *Shrubs, Trees and Flowers for Southern Africa.* McMillan, South Africa, 1975.

Fairchild Tropical Gardens. *Catalog of Plants.* American Horticulture Miami Society, 1979.

Graf, Alfred Byrd. *Exotica.* Roehrs, 1978. *Tropica.* Roehrs, 1978.

Hortus Third. By the Staff of the L. H. Bailey Hortorium, Cornell University. Macmillan, 1976.

Hoyt, Roland Stewart. *Check Lists for the Ornamental Plants of Subtropical Regions.* Livingston Press, 1938.

Libro del arbol, Tomo 1. Celulosa Argentina, 1976.

Libro del arbol, Tomo 2. Celulosa Argentina, 1976.

Mathias, Mildred E., ed. *Color for the Landscape: Flowering Plants for Subtropical Climates.* California Arboretum Foundation, Inc., 1973.

McClintock, Elizabeth, and Leiser, Andrew T. *An Annotated Checklist of Woody Ornamental Plants of California, Oregon, and Washington.* Regents of the University of California, 1979.

Morton, Julia F. *500 Plants of South Florida.* Fairchild Tropical Gardens, Miami, 1981.

Neal, Marie C. *In Gardens of Hawaii.* Bishop Museum Press, 1965.

Outdoor Circle. *Majesty, the Exceptional Trees of Hawaii.* 1982.

Sunset New Western Garden Book. By the Editors of Sunset Books and Sunset Magazine. Lane Publishing Co., 1979.

Synge, Patrick M. *The Royal Horticultural Society's Dictionary of Gardening,* 2nd Edition. Oxford University Press, 1969.

Thompson, H. Stuart. *Flowering Plants of the Riviera.* Longmans, Green & Co., 1914.